Science and Public Policy

Science and Public Policy: A Philosophical Introduction argues that in order to effectively apply science in any relevant or meaningful way, we must first understand what science is, how it works, and what its limitations are. The first half of the book thus discusses the domain of science, the concept of scientific evidence, and the interpretation of scientific data. The second half then moves through a detailed discussion of science communication in the public sphere, the concept, use and limitations of scientific expertise, and finally, the ways in which we can effectively apply science to public policies in the context of a democratic society. Along the way, the book uses detailed scientific examples to explore the relationship between science and uncertainty with the aim of showing that, in the end, public debates over science are rarely over the science itself, but instead over what public policies will follow from the science.

Key Features

- Explains what science is, as well as what it can and cannot do in the context of public policy
- Offers separate chapters dedicated to
  - Scientific Methodology
  - Scientific Evidence
  - Interpreting the Science
  - Communicating the Science
  - Scientific Expertise
  - Science-Informed Public Policies
- Includes Introductions and Summaries in each chapter to help guide the reader throughout the book

Ashley Graham Kennedy is Associate Professor of Philosophy in the Honors College of Florida Atlantic University. Her research focuses on applied issues in medicine, science, and global justice. She is the author of *Diagnosis* (2021).

Science and Public Policy
A Philosophical Introduction

Ashley Graham Kennedy

Routledge
Taylor & Francis Group

NEW YORK AND LONDON

Designed cover image: © shuoshu / Getty Images

First published 2024
by Routledge
605 Third Avenue, New York, NY 10158

and by Routledge
4 Park Square, Milton Park, Abingdon, Oxon, OX14 4RN

Routledge is an imprint of the Taylor & Francis Group, an informa business

© 2024 Taylor & Francis

ISBN: 978-1-032-31741-0 (hbk)
ISBN: 978-1-032-31740-3 (pbk)
ISBN: 978-1-003-31107-2 (ebk)

DOI: 10.4324/9781003311072

Typeset in Optima
by codeMantra

Contents

Acknowledgments

I would first like to thank the dean of my college, Justin Perry, for generously granting me a semester of research leave during the spring of 2022, which allowed me to complete this book according to schedule. Of course, I had no idea when I was granted this leave, nor when I signed the contract for the book, that during the writing of it I'd be hospitalized eight times for three life-threatening conditions, and ultimately undergo emergency open heart surgery to save my life. It is truly a miracle that I am even still here, and I am grateful for that! I am also grateful to Andy Beck, my editor at *Routledge*, for his genuine understanding and compassion during this time, as well as to all of my students at the Honors College of Florida Atlantic University for pushing me to think more deeply (and clearly) about the intertwined issues of science and public policy, and indeed for inspiring me to write these thoughts down in the first place. It is too often assumed by academics that research must always come first, and that teaching is somehow less important. I disagree. I have learned, and continue to learn, a great deal from my undergraduate students, and for this I am indebted and grateful to each one of them. I also appreciate the help of my colleagues Nicholas Baima, Marzia Marastoni, and Sarah Malanowski in discussing these issues with me, and helping me to clarify them, as well as to an anonymous reviewer, whose comments helped to make this book better than it would have been otherwise. And as always, I am so thankful for my family: my husband Bobby and my two sons, Micaiah and Kai. I truly do not know where I would be without you, and I love you more than words can say.

Preface

Background

The initial idea for this book was born from a lifelong love for science (from bug-collecting to stargazing, to balancing equations) coupled with a more recent, and growing, frustration with public policy, especially with the nature and tone of public debates over how scientific results should inform these policies. In following some of these debates, it became apparent that many in the public sphere did not seem to really understand what science is, or how it works, or what its limitations are. However, this problem didn't seem to be one of ignorance, rather, this lack of understanding is – surprisingly – equally apparent in non-scientist members of the general public, in policymakers, and even in practicing scientists as well. What is going on? There is certainly more than one contributing factor to this problem, however, one that stands out is the fact that while the STEM disciplines are widely appreciated, and of course taught at all major universities around the world, the philosophical foundations of science, on the other hand, are almost never taught at the undergraduate level, even to those who are studying to become scientists or policymakers, or both. Upon realizing this, it occurred to me that perhaps the significant and growing public gap in scientific understanding is not due to a lack of science education, but rather to a misunderstanding of what science is, as well as of what it can and cannot be expected to do. In fact, I was ultimately inspired to write this book because many of my own undergraduate students were often confused about the domain of science as well as its proper role in public policy formation – and this was no fault of their own. These students were, in some cases, taking advanced classes in neuroscience in preparation for PhD programs, or, in other cases, going on to study public policy or population health at the graduate level at highly ranked universities but yet they had never been taught about what science *is*, or what its limits are, or how we should use it to inform public policy. Such an understanding, I believe, will, in turn, facilitate better policymaking as well as more effective science communication. Thus, this book is an attempt at explaining both what science *is*, as well as what it can and cannot do, in a way that is accessible to bright undergraduate students. It is not, on the other hand, an attempt to replace or to compete with any of the excellent available works on values in science or the communication of science, or the roles of public trust and scientific expertise in public policy, many of which I draw on in the chapters that follow. Instead, my aim in this manuscript is to bring together the themes of the method, domain, and application of science, along the way discussing the issues of scientific communication, scientific uncertainty, and scientific dissent, in a manner that I hope is both interesting and readily accessible to students as well as to policymakers and other interested non-academic readers. Thus, I see this book as an introductory philosophy of science text for those who have a desire to understand the nature of science as well as the way in which it can be applied to public policy.

Book outline

In the first chapter of this book, I begin by outlining the domain as well as the method and the aims of modern science. To do this, I draw on an example from contemporary astrophysics. As I argue in the chapter, these things are vital to understand before proceeding to the further topics in the remainder of the book, because in order to effectively apply science in any relevant or meaningful way, we must first understand what it is, how it works, and what its limitations are. If we fail to do this, then we risk either over or underestimating its reach, which can negatively impact subsequent science-informed policy decisions. During the course of the discussion in this chapter, I also outline in detail the realist vs. anti-realist views of science, and the implications of these views for policymakers. I argue that whether we are realists about science or not, we can in fact trust the information that scientific models and theories give us, because the method of science, together with its in-built safeguards, has a historically reliable track record of success. In Chapter 2, I discuss in detail the concept of scientific evidence, focusing on what makes it different from other sorts of evidence. It is essential to have a conceptual understanding of the notion of scientific evidence, because it is sometimes misunderstood, even by scientists themselves, that there is no rule of logic that can decide for us the question of what counts as scientific evidence – or indeed as evidence in any given domain. However, this does not mean that weighing what should and should not count as scientific evidence is a hopeless endeavor, or one that lacks intellectual rigor or objectivity. Instead, we can conduct this process in an effective and rigorous way by examining both the scientific values (such as accuracy and predictive or explanatory power) and the extra-scientific values (such as pragmatic, social, and moral ones) that inform the scientific method itself. In Chapter 3, I discuss the interpretation of scientific data. As we will see in Chapter 1, modern science makes use of the methods of modeling, experimentation, and observation. Further, scientific models are often used in domains where observational experiments are difficult or impossible to perform. For example, epidemiology is a science that doesn't easily allow for direct experimentation, so epidemiologists often build models to understand and inquire into this area, as well as to make predictions regarding what public policy interventions might be effective in addressing community health problems that arise. But as we will see, modeling in the sciences also involves incorporating "place-holders," that is, idealizations or other assumptions, into the models that are built, and because of this, the data outputs of these models always require interpretation before they can be applied. Further, experiments, too, require decisions about what to include and what to control for, or even whether or not to run the experiment in the first place. Even the outputs of observational studies, because such studies are especially prone to interference from researcher bias, must be interpreted as well. What this means is that in practice *all scientific information requires interpretation* – no scientific data interprets itself. Thus, if we are to understand scientific information, we must first learn how to "convert" knowledge about the model, experiment, or observation to knowledge about the target system in the actual world that we are attempting to understand. Chapter 4 is a detailed discussion of science communication in the public sphere. Before science can be applied, it must first be communicated – to the general public, to policymakers, and/or to other relevant stakeholders. This task of communicating science to non-scientists, however, can at times be both difficult and complicated, because scientific results are often complex and sometimes therefore misunderstood. Further, in the scientific community, disagreement, in the context of research, is not uncommon, and this must often be communicated as well. However, as we will see in the chapter, such disagreement need not be a cause for concern, because when scientists disagree, or are uncertain about their hypotheses, models, or theories, this is often a good thing, as it can serve as motivation for further research. For instance, by building and comparing differing or "competing" models or hypotheses, scientists are often able to better understand the process being studied, by homing in on entities or processes that do not vary when modeling assumptions are changed. But in the public arena, scientific disagreement is sometimes misconstrued, or even exploited as, scientific controversy (see, for example, Cassam 2017; Havstad and Brown 2017; Goldenberg 2020). However, in the chapter I will show that disagreement and controversy, in the context of scientific research, are not the same thing and that scientific disagreement, on its own, should not prompt public distrust of either scientists or the methods that they use. In Chapter 5, I discuss the concept, as well as the implications and reach of scientific expertise. Most of us will recognize that it has become increasingly common to hear calls for the public to trust the "experts" or to "follow the science." The idea seems to be that scientific experts, over and above the general public, are able to tell us not only what to believe, but also what to do, in any given domain. However, as I will show in the chapter, there are several problems that arise from this overly simplistic view of the role of scientific expertise in the public domain. First, the definition of who counts as an expert is highly contested. Second, it is not clear that being an expert (alone) qualifies one to prescribe actions or policies. Given these points, I argue in the chapter, first, that determining what it means to be an "expert" and, in particular, a scientific expert, is important for the application of science to public policy because there is a relationship between scientific expertise and epistemic authority. Second, I argue that although we ought to trust scientific experts because we can learn from them what *is* the case, scientific expertise alone is not enough to tell us how we *ought* to act. Thus, this chapter explores the related questions of: who counts as an expert, why we ought to trust experts, and what is the normative reach of the information/knowledge that we can gain from experts in any given domain. Finally, in Chapter 6, I discuss the ways in which we can effectively apply science to public policies in the context of a democratic society, using the example of laws governing the use, prescription, and sale of addictive substances. With this example, I aim to show that the process of enacting science-informed public policies requires several steps, beginning with the determination of whether or not a given problem is worthy of a policy intervention, followed by the proposal of a potential solution, the communication of this solution to the relevant stakeholders, including both policymakers and the general public, and then finally, after implementing the solution, a determination of whether or not the intervention was, on the balance, effective and not harmful. Along the way, I explore in the chapter the relationship between science and uncertainty, with the aim of showing that, in the end, public debates over science are rarely over the science itself, but instead over what public policies will follow from the science.

January 2023, Tequesta, Florida

1   Scientific methodology

Scientific progress requires many methods.

(Blystone and Blodgett 2006)

Introduction

It is very likely that you picked up this book (or registered for a class that requires it) because you have an interest in both science and public policy, as well as the connections that exist, or that can be made, between the two fields. And this means that you already know that the findings of modern science, from the discovery that matter is energy, to the mapping of human DNA, as well as the many benefits of scientific discoveries, from life-saving medications to instantaneous communication across the planet, are simply astounding – sometimes even beyond belief. Scientific inquiry in the modern era has allowed us to make incredible discoveries, to better understand the world in which we live, and to live healthier, longer, and more productive lives in it. It's natural then, for us to also want to understand the role that science has, or should have, in the formation of public policies and governmental regulations. If science can reveal the hidden nature of reality, so to speak, shouldn't it also be able to tell us how to act in our world? And so, many of us have found ourselves wondering if and how science can guide, or inform, our decisions in the public arena. We will eventually get to this question. But before we can even begin to address it – before we can even begin to understand the potential ways in which science can and should influence policy decisions – we must first take a step backward, and begin by examining both what science *is* as well as what it can (and cannot) *do*. This will, in turn, involve examining the method of science, the domain of science, and the aims and limitations of scientific inquiry, as well as addressing the question of what counts as evidence toward a scientific theory (and who decides). These are the things we will discuss in the first half of this book. Then, building on this foundation, in the second

DOI: 10.4324/9781003311072-1

half of the book we will turn to the questions of how to interpret scientific results, how to make sense of scientific dissent and disagreement, how to communicate science to non-scientists, how to recognize scientific experts, and, finally, how to apply scientific results to public policy decisions in the context of a democratic society. In sum, the aim of this book is to first walk you through an understanding of what science is, and what it aims for, in order to prepare you to better understand how to apply it. The prior exercise is an important first step for anyone who is, or who eventually wants to become, a policymaker, because before science can be aptly applied to policy, its method, domain, and aims must be adequately understood. With this understanding in hand, we will then be able to recognize that while science alone cannot give us the answers to questions of morality, politics, or policy, it can, when effectively applied, definitely help to point us in the right direction.

The method of science

To begin, we should start by recognizing that science is not only a subject of study but is also a *methodology* that comes with a specified *domain* as well as a specific set of *starting assumptions*. In other words, as some philosophers have put it, science is not simply practiced in "thin air," so to speak, but rather begins with the presupposition of certain philosophical theses. This means that science, whether practicing scientists realize it or not (and they often do not!), cannot be done without first addressing certain philosophical questions. The most basic of these questions is, simply: "What is science?" On the one hand, this is an easy question to formulate, however it is a very difficult one to answer – as many others have previously noted (see, for instance, Potochnick 2017). Some thinkers have attempted to answer this question by pointing to specific examples, that is, by recounting a list of human activities or areas of inquiry that count as science – such as biology, physics, chemistry, botany, or economics. But when it comes to the task of providing an actual definition – a set of necessary and sufficient criteria – for what makes some activities a "science" and others not, then people begin to disagree. For instance, many of my own undergraduate students, when asked this question, begin by trying to define science, not by tying it to a domain, but instead by identifying it as any activity that conforms to the scientific *method*, which they usually describe as a method of experimentation. And while there is certainly something right about this – much of scientific inquiry is indeed governed by the experimental method – not all of it is. Astrophysics, which I studied as an undergraduate, for example, is an area of science which is not amenable to the traditional methods of scientific experimentation. Instead, astrophysics is a discipline that employs the methods of observation as well as modeling and computer simulations (two scientific techniques that will be discussed in more detail in Chapter 3). It does not (because it cannot), on the other hand, employ traditional experimental methods. We cannot, for example, set up a supernovae explosion or a black hole implosion or a planetary collapse in order to observe and study it, and this is because the experimental method involves intervening in nature in a way that this field does not allow for.

To further illustrate this point, as well as the basic components of the traditional experimental method, we can begin by describing it as a process that involves two test groups, one (the experimental group) which is intervened upon, and one (the "control" group) which is held constant. The idea is that such a system of intervention/control will allow us to identify what differences the intervention does or does not make to the entity, process, or system that is being studied, and thereby allow us to learn things about it. The simplest example of this method that comes to mind is the quintessential 4th grade science fair project that involves providing pea plants with water, soil, and sunlight in the control group but only soil and sunlight in the experimental group, and then recording the differences in plant growth rate and development between the two groups. This method (in various levels of complexity) is employed in many other areas of science as well, including chemistry, medicine, and physics. However, as we have seen, it is not used in all areas, such as in astrophysics, or epidemiology (which we will discuss in more detail in Chapter 3), where traditional experimentation methods are not possible, practical, or, in some cases, ethical. Yet, we would be hard pressed to find anyone who would argue that either astrophysics or epidemiology is not a science! So what makes something "a science," then, must depend not only on methodology but also on something else. This "something else" is what is known as the domain, or boundary limit, of science.

The domain of science

To delimit its domain, contemporary science makes use of the philosophical notion known as *methodological naturalism*. Methodological naturalism is the assumption that the *answers to scientific questions can only be given in terms of natural causes*. What counts as "natural," in turn, is defined by that which is encompassed by the domain of space, time, energy, and matter. While this domain is indeed broad, it is far from being all-inclusive. The concepts of philosophy, including ethics, for example, lie outside of this domain. In other words, philosophical concepts, such as justice, fairness, good, and bad are not the sort of thing that can be understood via empirical methods, which are the methods that we use to investigate the realm of the natural world (that which is encompassed by space, time, energy, and matter). The same is true of the concepts employed in public policy, legal theory, and political and moral theorizing more generally. What this means in practice is that while science is an impressively reliable method, it is limited by its domain, and because of this it is not a method that is able to provide the answer to every question, nor the solution to every problem: science alone cannot tell us, for example, whether or not there is a supreme being or whether or not genocide is wrong, or how we could go about stopping it if it is. It cannot tell us these things because these questions lie outside of its domain of inquiry. However, we should not understand this to be a flaw of the method of science. Instead, it is an *intentional* feature of its design, which means that these questions lie outside of the domain of scientific inquiry purposefully. Thus while science itself takes no position on the *existence* of supernatural entities, processes, or concepts (in the strict sense of something that is outside of space, time, energy, and matter), it restricts appealing to anything that is outside of the natural within the context of scientific inquiry, research, experimentation, or explanation. In other words, in the domain of scientific inquiry, all explanations are required to be *naturalistic,* in that they must not appeal to anything outside of the domain of the natural world. This is why science, on its own, cannot give us the answers to ethical, social, political, or theological questions. It cannot do so because it was not intended to. Instead, the answers to these questions must always appeal to non-scientific concepts or principles of one kind or another. For example, in making policy decisions we might want to consider ideas of equity, justice, well-being, or fairness. But none of these are concepts that are amenable to scientific inquiry, which is by definition empirical. What this means is that we cannot learn about these concepts via our five senses or their extensions: we cannot learn what justice is by looking through a microscope or a telescope, for example, in the way that we can learn what a proton or a quasar is. Instead, justice and equity are non-scientific, philosophical concepts. So it is not possible, even in principle, to design an experiment that could tell us what, for example, justice is. In other words, we cannot learn what justice is by simply using our senses to observe the natural world – the world which is delimited by space, time, energy, and matter. Instead, an analysis of the concept of justice, and indeed of all non-scientific concepts, requires some level of non-empirical analysis. And this kind of analysis, by design, lies outside of the domain of scientific inquiry.

The aims of science

Related to the question of the domain of science is the question of what scientific inquiry aims for, or, to put it another way, the question of what the overarching goal of science is. For many, the first thing that comes to mind regarding this question is that the goal of science is to reveal facts, or truths, *about the natural world*. This idea that the goal of science is to help us to learn about the way that the natural world "really is" dates back to at least the time of the Enlightenment thinkers (Schmidt 1996). More specifically, it is the view that all scientific knowledge is derived from sensory experience and, when conducted by disinterested observers, is not subject to any sort of systematic error or bias. In other words, on this view, science is supposed to tell us about the way the world really is, and not the way that we merely think it is, or want it to be. As we will see in what follows, this view of science, which is known as the realist view, has since been challenged in many ways. However, even in spite of these challenges, scientific realism is still very much alive and well, and is the view that is adhered to by many natural scientists today (Psillos 1999; Beebe and Dellsen 2020). To further elaborate, the realist view of science holds that the primary aim of all scientific inquiry is to provide realistic explanations as well as accurate predictions of natural phenomena, both via ideas that are testable by empirical methods. To be more specific, there are two main forms of scientific realism. The first is what is known as epistemic realism, which is the view that our best scientific theories are approximately true, or at least *aim* at the truth, even if they never quite reach it. This, in turn, means that our scientific theories can be understood to give us knowledge about the actually existent, mind-independent world – a world that exists in a certain way, no matter what anyone thinks. On this view, scientific reality is not something that we do (or even could) create – rather it is the view that, even if there were no humans to investigate it, the world would still be the way that it is. The second form of scientific realism is what is known as metaphysical realism, which is the view that the entities (whether those that are observable either directly via the senses or by their extensions, such as microscopes or telescopes, or those that not observable at all) that are postulated by our best scientific theories actually do exist. In other words, metaphysical realism is the view that there (a) exists a mind-independent natural world and (b) that we can gain knowledge about this world via scientific inquiry. This knowledge criterion of scientific realism has, built into it, the notion of objective truth, which is a notion that we will appeal to repeatedly in the chapters that follow. An objective truth is defined as one that does not in any way depend upon the person (or subject) who espouses it. Further, it is, in general, accepted by philosophers that in order for someone to know something at all, then that something must be true. That is, if we have a particular belief, scientific or otherwise, that is false, then that belief does not count as an instance of knowledge. For example, if I believe that the earth is flat, this does not count as an instance of knowledge if it is false that the earth is flat. It is important to note that both epistemic realism and metaphysical realism have built into them this objective truth criterion.

Further, if we put epistemic realism and metaphysical realism together, this, in turn, gives us the strongest form of scientific realism, which is the view that our best scientific theories give us either true or approximately true descriptions and predictions of both the observable and unobservable aspects of an actually existing mind-independent natural world. Here we should note that while scientific realists, of any sort, do not claim that *all* scientific theories are true, or that the ones that are true give us any sort of predictive or explanatory *certainty,* they do, in general, make the claim that scientific knowledge is something that *progresses over time*. In other words, the realist claims that as time goes on, we are gaining more, and more precise, scientific knowledge, that is, that we are continually learning more about the world around us.

Now that we better understand what scientific realism is, what can be said in support of this view? Perhaps the most commonly advanced argument in favor of scientific realism is the one which is known as the "no miracles argument." The claim of this argument is that realism is the only view that doesn't make the historical success of science a "miracle." To be more precise, the no miracles argument can be formulated in the following way:

Premise 1: Science has a track record of impressive success.
Premise 2: This success is best explained via scientific realism.
Conclusion: Therefore, scientific realism is the correct view of science.

There are, however, several potential ways to object to this line of argument. One such objection is known as the underdetermination of theory by data: because, the objection goes, in many cases scientific data can be explained, and accurate predictions can be made, on the basis of more than one theory, this calls into question which one of the potential theories is the correct one. And if we don't know which one of two or more theories is correct, then we don't have knowledge (in that instance) about the part of the world in question.

A second objection to scientific realism is what is known as the pessimistic induction. The pessimistic induction is the worry that because there has, historically speaking, been a regular turnover of older theories for newer ones, this means that from the point of view of the present, most past theories must be considered to be false. Generalizing inductively from these cases, this, in turn, means that the scientific theories that are accepted at any given time will, very likely, ultimately be replaced and regarded as false from some future perspective. Therefore, the argument goes, our current scientific theories are also very likely to be false.

These worries, amongst others, have led some to adopt an anti-realist view concerning science. As with scientific realism, there are also several forms of scientific anti-realism. The first is what is known as empiricism, which is the view that all scientific knowledge comes only from observation and, therefore, we cannot have theoretical knowledge of any unobservable phenomena (such as quarks, or electrons, etc.) In particular, one popular form of empiricism is known as instrumentalism. This is the view that scientific theories are simply instruments for predicting observable phenomena and do not give us – or even aim to give us – accurate explanations about phenomena in the natural world. For example, modern empiricism, as outlined by van Fraassen (1980) argues that the aim of science is not truth, or even approximate truth, but instead is only empirical adequacy, where "a theory is empirically adequate exactly if what it says about the observable things and events in the world, is true" (van Fraassen 1980). This view is antirealist because it recommends belief in our best theories only insofar as they describe observable phenomena, and an agnostic attitude with respect to anything that is not observable. Because of this, some understand this view to be a form of "partial" scientific realism.

A second form of scientific anti-realism is what is known as historicism. This view, which was advanced by Kuhn (1962), says that when we look carefully at the history of science, we can see a recurring pattern: there are periods of what he called "normal science," which are often fairly long in duration (such as, for example, the period dominated by classical physics), which are then, in turn, punctuated by "revolutions," which eventually lead scientific communities from one period of normal science into another, via a rapid and all-encompassing change of (rather than a slow progression to a new) paradigm. On this view, then, science does not actually make progress in the way that the realist claims it does. Instead, for Kuhn, there is no scientific progress, only change. For example, he argues that the shift from Newtonian physics to general relativity did not occur because we learned something new, but rather because of a complete and revolutionary change in both perspective and understanding. While Kuhn admits that Newtonian physics and general relativity are different, on his view, what examples like this show us is that each paradigm on either side of a revolution such as this one requires a commitment to an entirely different world view, and thus the two paradigms simply cannot be compared with one another in any meaningful way. This, in turn, means that, on the historicist view, we have no basis for the claim that scientific knowledge progresses over time, even if it does change.

A third sort of anti-realist epistemology of science is what is known as constructivism, which is the view that science is never a purely objective enterprise, but instead is one that is always influenced by social factors as well as by extra-scientific values. Thus even though "a commitment to studying the sciences from a sociological perspective is interpretable in such a way as to be neutral with respect to realism" ("Scientific Realism," Stanford Encyclopedia of Philosophy) In practice, most constructivist accounts of science are either implicitly or explicitly antirealist and adhere to the view that human values both inform and influence the production of all scientific knowledge, which, in turn, means that scientific knowledge is always open to the possibility of systematic errors or biases.

Finally, a fourth way of viewing the aims of science, and one which crosses the traditional realist/anti-realist divide, is what is known as social empiricism. This view is advanced by Solomon (2001). Solomon's epistemology of science is also a social account of scientific rationality, but it differs slightly from constructivism in that it is not inherently anti-realist, but instead proposes that the pursuit of truth in scientific research can be consistent with an incorporation of extra-scientific values. Solomon further argues that while scientific realists and anti-realists disagree on many points, they actually agree on a shared set of assumptions about the nature of rationality and progress in science, and that, because of this, we need not be as skeptical of scientific progress as anti-realists have argued that we should be, even when we allow that social factors do, in fact, often influence scientific inquiry.

In sum, while we can clearly see that there is significant disagreement between realists and anti-realists over what constitute the aims of science, and, in particular, over whether or not science aims at "truth" in one form or another, there is also widespread agreement amongst both camps that scientific inquiry strives for *objectivity*. That is, even those who argue that science is, and even should be, influenced by human interests and values that fall outside of the domain of scientific inquiry, still recognize that scientific theories can be considered to be good ones only insofar as they aim either for some form of realistic explanation or for some form of empirical adequacy, rather than relying on personal whims or desires. Thus, both scientific realists and scientific anti-realists agree that if the predictions made by a theory are inaccurate, then that theory is not a good one.

However, it must still be acknowledged, by realists and anti-realists alike, that whether the aim of science is truth or merely empirical adequacy, that one of the primary tools of modern science, that of scientific modeling, presents us with a conundrum of sorts, because models in science always incorporate idealizations, approximations, simplifications, abstractions, or other kinds of inaccurate components. Yet, scientific models are used to give us (at the least) accurate predictions and (hopefully) adequate explanations of empirical phenomena. That is, they are thought to be tools that are capable of providing us with knowledge about the natural world. An examination of the practice of scientific modeling thus shows us that, on the one hand, we need not be realists about science in order to effectively apply its methods, but that, on the other hand, we need not be anti-realists, either, even upon the realization that no scientific model is ever a completely accurate representation of its target system in the natural world. To see this point in more detail, consider the following example.

Example: astrophysical model

As an undergraduate, I was part of a team that worked on a model of a comet (Graham et al. 2000) that was built in order to determine the likelihood of observing a cometary maser during a random ground-based telescope search. The term "maser" stands for microwave amplification by stimulation emission of radiation. (A laser is a maser that works with higher frequency photons in the ultraviolet or visible light spectrum.) Specifically, we built this model because previous attempts to observe a maser in the comet (Hale-Bopp) that we were studying had been unsuccessful. This negative result was surprising to us because water maser emission had been observed in other comets that had a much lower water production rate. With the model that we built, we wanted to explain this specific negative result in Hale-Bopp, but we also wanted to understand more generally what conditions are required in order to successfully observe a cometary maser from the ground. To do this, we built a model that incorporated previously available observational data, along with several idealized assumptions. First, in our model, we assumed that gas sublimation from the surface of comets is not uniform, but takes place only in jets, and that therefore, masing from any cometary surface also takes place only in jets. This first assumption was, strictly speaking, not true, and we, of course, knew this. Although the surfaces of comets are inhomogenous and contain both more volatile regions ("hot spots") and less volatile regions, and therefore it is likely that gas sublimation and masing would sometimes take the form of jets, it is not likely that they would *always* take this form. Yet, as we will see, this assumption that we made in our model turned out to be a useful one. Second, in our model we also assumed that every comet has only one jet (which we also knew was not true, given the previously available observational data), and finally, in the model we assumed that a cometary jet is always located either near the pole or near the equator of a rotating comet (again we knew this to be an idealization, but incorporated the assumption to simplify the model). This third assumption is illustrated in Figure 1.1.

In addition to these idealized assumptions, in the model we also accurately represented several previously known components of a masing comet including the chemical makeup, the masing mechanism (collisional de-excitation in the expanding gas that emanates from the comet's surface), and the minimum temperature at which a maser can be detected (approximately 10^6 Kelvin).

Our model, with both its accurate and inaccurate components and inputs, allowed us to explain a previously unexplained aspect of the specific target system (comet Hale-Bopp) and it also explained the necessary conditions for cometary maser observation, in general, in the following way. First, the model incorporated idealized information as described above. These inaccurate, purposely idealized, assumptions were initially included in order to

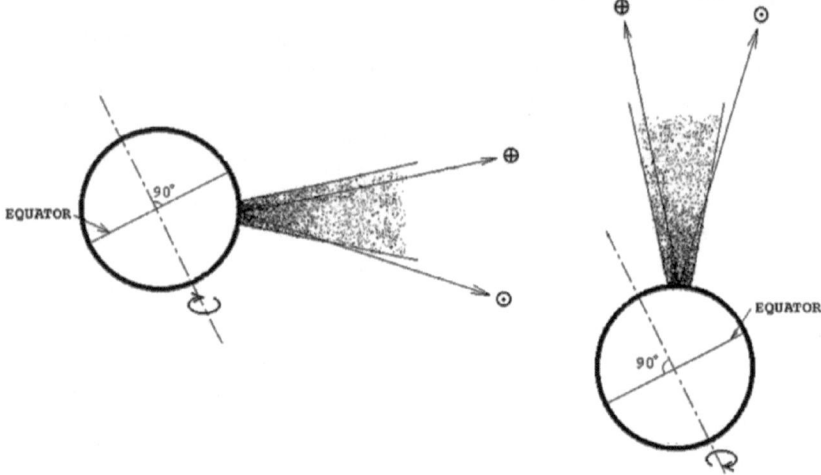

Figure 1.1 The diagram on the left shows a comet with a masing jet located near the equator of a rotating comet. The diagram on the right shows a comet with a masing jet located near the pole of a rotating comet.

make the model simple[1] enough for calculational purposes. However, even though the assumptions that went into our model were originally chosen for a pragmatic purpose, they turned out to have a non-pragmatic, explanatory benefit as well. This is often the case in scientific models. Yet while it is widely recognized that scientists include idealizing assumptions in their models for pragmatic reasons, it is not as widely recognized that the benefit of these assumptions often extends far beyond pragmatics. However, in an examination of how to apply science to policy (as we will see in more detail in later chapters), this is an important point to understand. We need to recognize that idealization, or the practice of making deliberately simplifying assumptions when conducting scientific modeling, is not only non-harmful, but is actually often beneficial – a point that we will return to when we discuss the application of scientific modeling to public policy in Chapter 3. For example, friction and air resistance are often given a zero value in simple physics models, not just to make the models easier to work with, but also because friction and air resistance, in many cases, are not relevant to whatever it is that is being explained by the model.

But there is another, more active, role that idealizations often play in model explanation. For instance, in this particular example, because our comet model was simplified by placing the cometary jets at the pole and the equator of the comet, it allowed us to make the calculations necessary to make probability estimates about the likelihood of observing a cometary maser in a random search from the ground. In our model, we gave the jets strategic (although not actual) locations. We could not have put the jets just anywhere and have gotten the same results. Instead, we had to put the jets at the two possible extremes in order to ensure that the actual case, whatever it turned out to be, would be encompassed within the model. (This sort of strategic idealization is also used in social science models, as we will see in later examples in this book). Thus, our model had explanatory power in part *because of* the idealized assumptions that it employed. And again, as we will see later in more detail, this is often the case with scientific models: sometimes the idealized components can actually help to facilitate more accurate predictions or explanations.

The explanatory success of this model also highlights the importance of idealizations for model explanation more generally – it is not always the case that the more realistic a model is, the better (in terms of either predictive or explanatory power) it is. Thus while, in general, explanations do not *require* idealizations, neither does idealization rule out genuine explanation.[2] However, at first glance, this does seem to present a problem:

1 Scientific models are (strictly speaking) inaccurate because they misrepresent their targets via various idealizations.
2 Scientific models are nevertheless (in many cases) explanatory.
3 Only true accounts can give accurate explanations.

One way to solve this problem is by rejecting the third premise. But this then gives rise to a new issue. The worry is then that it isn't clear how a model that makes use of idealized components can *really* explain an actual target. However, if we understand a realistic explanation as one that tells us something about the way the world really is, or the way the world really behaves, then we can better understand how idealized models can do this: they do so by telling us something about the actual properties of or processes within the modeled target even if they employ idealized assumptions in order to do so. For example, the comet model that we built delivered an accurate explanation of the situation because it conveyed information about the geometry that is required for a maser to be observed from the ground. That is, it identified the property – a geometrical configuration – that is required for successful detection of these phenomena. To put the point more generally, because idealized components in models do have real-world correlates, the models that employ them are able to tell us something about the way the target really is. Thus, what this means is that realistic explanation does not require complete and accurate representation of the target.

One might wonder what, if anything at all, the preceding discussion of scientific modeling has to do with the central question of this book, which is that of how to effectively apply science to public policy. After all, astrophysics isn't usually the science we are thinking of when we think about public policy. However, because physics is considered by many to be the quintessential "hard" science – that is, the sort of science that we think of when we think of one that gives us pure "facts" or unbiased truth about the natural world – this example serves to illustrate the point that *all* models incorporate idealizations. This is not done because scientists are devious or have some sort of hidden agenda (we will return to this discussion when we talk about scientific disagreement in Chapter 4) but rather because it is a method that works. Further, the discussion of this example was also meant to illustrate that scientific models, which are one of the primary tools of the modern scientific method in almost every area of science, need not be 100% accurate (because indeed they never are) in order to either accurately predict or adequately explain their target systems in the natural world. This, in turn, is relevant for the application of science to policy: it tells us that it is not only OK but also to be expected that scientific models will not ever be 100% true, but yet this does not mean that we cannot (a) learn from them and (b) trust their results to be reliable and informative. Thus, whether we are scientific realists or scientific anti-realists, a wholesale distrust of scientific methodology, and thereby of scientific results, is not justified. Such skepticism would be unjustified because we have good evidence that the method of science is, in fact, a very reliable one as evidenced by the historical track record of science. This historical record shows us, whether we are realists or anti-realists, that science has a long history of resounding successes in terms of both predictions and explanations of natural phenomena. This means that we can trust that the method of science will reliably give us (at least reasonably) accurate information about the world around us. This does not (as we will see in more detail in later chapters) mean that science gives us certainty: it does not. It also does not mean that science necessarily gives us truth, as that claim, as we have already discussed, is up for debate. Yet, because we are able to point to countless examples that show that the method of science does in fact give us accuracy, we know that we can trust it. And this, in turn, is what is important for, and relevant to, the questions of the application of science to public policy that we will address in later chapters. Ultimately, what we need to know in order to effectively apply science to policy is that we can trust the method of science to give us accurate information. In other words, we need to know that the method of science is a reliable one, and this is exactly what the long history of scientific success tells us.

Conclusion

In this chapter, we have examined the methods (modeling, experimentation, and observation), domain (space, time, energy, and matter), and aims (either truth or empirical adequacy) of modern science. Along the way, we have also discussed the realist vs. anti-realist views of science, and their implications for both practicing scientists and policymakers. We have seen that both scientific realists and anti-realists incorporate a notion of objectivity into their conception of scientific inquiry, even though they differ on what exactly that means. We have also seen that it is not necessary for a model (or an experiment) to provide us with perfectly accurate data in order for its results to be reliable and thus for the model to be trusted as a source of knowledge about the part of the world that it represents – knowledge that can eventually be effectively applied to policy decisions. Finally, the discussion in this chapter was also meant to show that it is important to have a general understanding of these foundational topics in science before we can be prepared to ask the question of how to best apply scientific data public policy. If we do not understand what science is, how it works, what it aims for, or what its limitations are, we will either over or underestimate its predictive and/or explanatory power and thus will not be able to effectively apply it to our policies in any meaningful or helpful way, or to accurately communicate the results derived from it. Given this, before we turn to the question of scientific application in the second half of the book, we will next discuss, in the following chapter, the concept of scientific evidence, taking into consideration the question of what counts as evidence in the scientific domain, as well as the question of who decides, and how. Then, in Chapter 3, we will turn to the question of how to interpret, and make sense of, scientific data and results, before discussing the topics of scientific communication in Chapter 4, and scientific expertise in Chapter 5, and then finally, a specific example of the step by step process of how scientific data can and should be applied to public policy, in Chapter 6.

Notes

1 Not all idealizations involve simplification. Instead, an idealization can be understood as a deliberate assignment of convenient yet unrealistic interpretations to real world entities, properties, or processes. Often these interpretations are simpler than their actual real-world counterparts, but this is not always the case.
2 This issue is discussed in more detail in Chapter 3.

2   Scientific evidence

Evidence is a piece of information that supports a conclusion.[1]

Introduction: what is evidence?

In addition to examining the method, domain, and aim of science, the concept of evidence is, perhaps not surprisingly, also of pivotal importance to an understanding of the modern scientific endeavor. And although evidence can be a notoriously difficult concept to define, one helpful way to think of it, and the way that I will define it for our purposes here in this book, is as *reasons,* or justification, for a particular belief (scientific or otherwise). As Kim (1988) argues:

> the concept of evidence is inseparable from that of justification. When we talk of 'evidence' in an epistemological sense we are talking about justification: one thing is 'evidence' for another just in case the first tends to enhance the reasonableness or justification of the second.

Further, different types of beliefs are justified by appealing to differing types, or categories, of evidence. For example, consider the scientific belief, or assertion, that "the sun is the center of the solar system." This statement is justified via an appeal to a different sort of evidence than is the moral belief/assertion that "genocide is wrong." But this is *not* to say that one of these two beliefs is stronger or more objective, or better supported than the other, as many often mistakenly assume. To be more specific, it is *not* the case that scientific beliefs are somehow better supported by evidence or are "more objective" than moral ones. And further, the difference between these two kinds of claims is not that one is fact and the other is opinion (which, as we will see in a later chapter, is a false distinction anyhow) – or that one is true and the other has no truth value at all. Instead, the difference between these two types of claims is that one is empirical and the other is normative. An empirical claim is a claim about what *is* the case, or

DOI: 10.4324/9781003311072-2

the way that the world is, while a normative claim is a claim about what *ought* to be the case or about the way that things ought to be. Both of these types of claims can be backed by equally strong reasons, but the reasons will necessarily be of differing kinds. What this means, in the context of our discussion of evidence, is that these two types of beliefs are justified by appealing to two different categories of evidence – in this case to scientific evidence vs. moral (normative) evidence, respectively.

The two examples I just gave, the claims that "the sun is the center of the solar system" and "genocide is wrong," might seem to be overly simplistic, precisely because in the actual world, many, or perhaps even most, beliefs or assertions, do not clearly fall into one category only but are rather instead what we might think of as "mixed" beliefs – in that they are comprised of both empirical and non-empirical components. As we will see throughout the examples that follow later in this book, this is particularly the case with many claims that are used to inform public policy. For now, let us take a look, as an initial example, at the claim that "all school children should receive the polio vaccine." This is the kind of claim, that if one were to make it, would need to be justified by an appeal to both scientific data (such as the information that the polio vaccine is indeed both safe and effective at preventing infection with the polio virus) as well as moral principles (such as the idea that preventing communicable illness in a human population is a desirable thing to do). There is no way to justify a belief/assertion like this based on scientific data alone because the assertion itself is not a purely scientific one. Instead, as we will see in further detail in Chapters 4 and 5, in cases like this, we must make an appeal to social, political, and moral values, particularly when our aim is to apply science to public policy, because all such application extends beyond the realm of pure science. This means that, in spite of the way that some policymakers talk, there is no such thing as simply "following the science" when it comes to this important endeavor. As we have already seen, science is indeed a reliable method, but this does not mean that it is a simple or straightforward prescription for public action. In other words, while science can tell us what is the case (in the form of empirical claims), it cannot tell us what to do with that information. When it comes to public policy, therefore, we must be prepared to appeal to extra-scientific concepts and values in our decision-making processes. Thus, the underlying point to take away from this discussion is that in the case of all beliefs – including scientific, non-scientific, and mixed ones alike, we should understand a reasonable one to be an assertion that is adequately supported by the right kind (or kinds), of evidence, rather than one that is of a given type or that falls within a certain domain: scientific beliefs are neither better nor stronger than moral ones, and indeed we must remember that in order to aptly apply scientific beliefs to public policy, we will ultimately need to incorporate concepts and values that fall outside of the domain of science.

Another important point to understand about evidence is that, in principle, evidence can be either subjective, in that it depends on the subject or person who holds it, or objective, in that it does not.[2] When it comes to supporting our beliefs, especially scientific and moral ones, objective evidence is considered to be preferable to, and stronger than, subjective evidence, because it is less likely to be dependent upon personal whims, preferences, or desires. For example, it is my personal (and deeply held!) belief that dark chocolate is superior in taste to milk chocolate. My reasons for this are, however, inherently and necessarily subjective – they depend solely upon my personal taste preferences and not upon anything else. And, clearly, others have tastes that differ from mine, and, worse than that (!), there is no way to objectively weigh whether or not dark or milk chocolate is the tastier option: when it comes to personal preferences there simply is no objectively right answer. However, when it comes to scientific, moral, or mixed beliefs, this is not the case at all, and we can see why objective evidence should be preferred over subjective. If it is important for us to know, for example, whether or not genocide is wrong, then we should *not* depend on anyone's personal preferences or tastes to answer this question. Rather, we need to collectively decide upon the objective reasons that support our holding of this belief (or not). The same goes for scientific beliefs: if we are going to assert, for example, that the polio vaccine is safe and effective, then we need to support this claim with objective reasons rather than subject-dependent ones.

Even once we have a general idea of what evidence is (reasons for belief), and why we need it (in order to justify our claims and beliefs to ourselves and others) – we still have to set ourselves to the task of determining what exactly should count as evidence toward a specific belief or claim in any given domain. Since this is book on science and public policy, we, of course, will be concerned with what counts as evidence for beliefs that fall within these two domains. In the previous chapter, we saw that the domain of science is limited to that which is contained within space, time, energy, and matter. This means that a moral reason, for example, given in support of the claim that "the sun is the center of the solar system," really would not be a reason in favor of the belief at all, since the claim is a scientific one, and any moral reason given in support of this claim would fall outside of the domain of science and therefore would not apply to it. However, in the case of a mixed belief or claim, such as the claim that "all school children should receive the polio vaccine" we will, as we have already seen, need to appeal to more than one kind of evidence in order to justify the assertion. In this particular case, reasons given for the justification of this claim would need to include both scientific and extra-scientific ones. And this is always the case when it comes to claims regarding the application of science to public policy – these kinds of claims require both types of supporting reasons.

How do we gather scientific evidence?

Now that we know what evidence is, how do we go about finding it? As we already saw in Chapter 1, one of the primary methods of gathering scientific evidence is via experimentation. There are many different ways of conducting scientific experiments, but one that is particularly common in both contemporary medicine and the social sciences, such as economics, and political science, is what is known as the randomized controlled trial. A randomized controlled trial (RCT) is an experiment that is controlled in two specific ways. Under the first control, the participants in an RCT are divided into two groups (via random distribution): the active intervention group, and the control, or placebo, group. Distributing participants into the two experimental groups in this way is meant to (hopefully) evenly distribute any potential confounding factors (which are factors that would contaminate, or make unreliable, the data from the experiment) between the two groups. The idea is to ensure that the results of the experiment are valid, or at least as reliable as is possible, given real world constraints. However, to mathematically *guarantee* that randomization would evenly distribute all known or unknown confounding factors between any two groups, an infinite sample size would be needed, and this is, of course, never the case in actual practice. Thus, the best that we can hope for is that randomization in our experimental trials will do *something* positive when it comes to the distribution of factors that we are not aware of. (For factors that we are aware of, randomization is not necessary for even distribution, this can instead be done manually, or via trial inclusion and exclusion criteria.)

The second way in which an RCT is controlled is by what is known as the process of double-blinding. Properly executed double-blinding means that both the participants and the researchers in the trial are "blind" in that they do not know who is in which group of the trial. This aspect of the controlled trial is supposed to mitigate researcher and/or participant biases that might potentially influence the results. For example, in trials of medical interventions, patients who know that they are receiving the intervention often fare better than those who know (or who think that they know) that they are not, while those who think they are not receiving the intervention often fare worse (even when they, in fact, are receiving it!) (Holman 2015). Further, researchers tend to (unconsciously) treat those who are in the intervention group differently from those who are not, thus potentially confounding experimental results, and blinding is meant to block this sort of situation.

In sum, because of these two specific types of experimental controls, RCTs are considered by their proponents to be much stronger sources of evidence than case studies, which are observational experiments that do not employ either of these controls. However, historically, before the use of RCTs began to become more widespread, case studies *were* considered to be good sources of scientific evidence (and as we will see in the following chapter, some still believe them to be). It was not until the late 1990s, when Evidence-Based Medicine (EBM) became popular as a new movement in medicine and solidified the reified status of the RCT in both scientific and medical research, that this idea of case studies as evidence began to be challenged in a widespread way. EBM, when it was initially proposed (and EBP – or "evidence based practice," as it is often now called), was different from the previous scientific paradigm in that it set forth a new technical definition of the term "evidence," as well as a hierarchy which defines this term:

> At the top of this hierarchy are randomized controlled trials (RCTs) and metanalyses, which are considered to be the strongest form of evidence, followed by observational studies, case studies, mechanistic reasoning and, finally, at the bottom of the hierarchy, is expert judgment or clinical expertise, which is considered to be the lowest form of medical evidence. The hierarchy of evidence is arranged in this way with the goal of minimizing bias, or systematic error, when gathering evidence, or pursing [scientific] knowledge. The idea is that randomized, blinded trials are less amenable to bias from either allocation or expectation than are observational studies. Further, it is thought that [researcher] expertise might be especially prone to being clouded by biases of various sorts, including expectation biases and interpretation biases that might be based upon incorrect or inadequate theories, and thus this form of evidence is ranked the very lowest on the pyramid. Overall, the main emphasis in the proposed method of EBM is on experimental data over and above theory. That is, although all data requires interpretation, and therefore, theory, at least at some level, when experiment and theory contradict one another, EBM tells us to rely on experiment (and revise the theory) rather than vice versa.
>
> (Kennedy 2021)

Thus although both medicine and the social sciences appealed to "evidence" long before EBP came onto the scene, the new paradigm proposed a new understanding of what the term "evidence" is meant to refer to – under the new paradigm it is meant to refer to experimental data rather than theory, observation or expertise of any kind. Thus, according to EBP, randomized controlled trials are the strongest form of scientific evidence that there is, and, according to some, the only sort we should use in our science-based decision-making processes. Essentially what this means is that, in practice, when a proponent of EBP uses the term "evidence," he or she is nearly always referring *only* to RCT evidence, and not to anything else.

Given the strength of this EBP evidence claim, that only RCTs should count as evidence for decision-making, one might reasonably ask how we actually *know* that RCTs provide us with the best form of scientific evidence available? The answer to this question, which might come as a surprise to some readers, is that according to EBP's own lights, we *don't* actually know this. In other words, we don't know that RCTs are stronger than other forms of potential scientific evidence because the only justification we have for the claim that they are is that expert opinion tells us so (Sehon and Stanley 2003; Howick 2011). In other words, EBP experts say that this is the case, and while they might in fact be right, according to the EBP evidence paradigm itself, expert opinion is the lowest form of evidence that there is (because it is subject to multiple forms of bias), and therefore is essentially something we ought not to trust. So according to EBP, we don't actually have any strong, objective evidence for its central claim that RCTs should be at the top of the evidence pyramid.

Where, then, does this leave us? First, we should note that although we do not have any *objective* evidence that supports the claim that RCTs provide the strongest form of scientific evidence, this does not mean that the claim itself is wrong. Instead, it simply underscores the point that EBP, and indeed science itself is a reliable, yet fallible, method. More particularly, it shows us that it is not possible for an RCT to answer *every* scientific question, or to provide perfect data in any given case. What this, in turn, means is that we will often need to appeal to other forms of evidence, such as observational evidence, theory, or expert opinion, when we are weighing in on important questions in science and public policy. Yet, those who strongly adhere to the EBP paradigm often speak as if this is unnecessary, and that appeals to theory, expert opinion, or observation are not helpful and sometimes even harmful. However, this overly constrained view of what counts as scientific evidence does not hold up under closer scrutiny. For example, no RCT for penicillin as a treatment for strep throat has ever been conducted. But does that mean we have *no* evidence that this therapy works? Certainly not. We do, in fact, have very good evidence that penicillin works to effectively treat throat infections with streptococcus A bacteria. But the evidence that we have for this view is not experimental – it is observational, in the form of many, many case studies. That is, we have simply, yet clearly, observed that countless cases of strep throat have been cured via treatment with this antibiotic, and although the evidence we have for this is indeed observational, it is not subjective – everyone agrees on these cases of cure.

Examples such as this one force us to choose: should we be strict adherents to EBP and hold to the claim that only RCTs count as strong evidence for scientific beliefs? Or should we allow ourselves to be pluralists about evidence, at times incorporating observation and/or theory into our reasoning processes? In what follows, I will argue for the latter, both because it makes for better science and because it makes for better science-based policies.

Before we take a look at an example that helps to illustrate this, we should first examine one further issue that arises with the use and application of randomized controlled trials in the context of scientific experimentation. This issue is what is known as the problem of generalization. Simply stated, the problem is that in any given (well-designed) RCT, there are inclusion/ exclusion criteria which tightly restrict the profile of the participants in the trial, in an effort to increase trial validity by reducing the number of known confounders. This means that, often, a trial population looks very little like the target population in the actual world that it is meant to address, and that the results of the trial are meant to apply to. For example, in many RCTs of medical interventions, participants with more than one medical condition, which are known as "comorbidities," are excluded from participating in the trial, in the interest of controlling for both known and unknown confounding factors. However, in the general population, which is any population that goes beyond that of the trial, many people (who will ultimately be given the intervention being tested) do, in fact, have more than one medical condition, and this often affects whether or not the proposed intervention will be safe and/or effective to use in these particular people. And this is the sort of situation that should cause us to worry about how and whether we can apply the results of any given RCT to an actual target population, because in order to do so, we need to know that the trial population and the target population are alike in all of the relevant ways. But this is pretty much impossible in practice. Although a more detailed discussion of and potential solutions to this problem of generalization is beyond the scope of this chapter, the important point for our purposes is that we should recognize that in many instances, it may be necessary to supplement our RCT data with observational information and/or other forms of evidence before applying the data outside of the trial domain. Thus, we ought to be skeptical about the supremacy/self-sufficiency of RCTs on their own, even if they are often excellent sources of scientific evidence. Instead, we should recognize that the results of RCTs often require interpretation within the context of the consideration of other forms of non-experimental evidence, such as observation and theory, as well as extra-scientific values. The following example will help to further illuminate this point.

Example: polio vaccine research

The polio vaccine is widely considered to be one of the safest and most effective vaccines ever developed. It was created in the 1950s in order to prevent poliomyelitis, a highly contagious and debilitating enteroviral disease that spreads from person to person via direct or indirect contact and can cause paralysis, disability, and even death. According to the World Health Organization:

> Once viral reproduction is established in the mucosal surfaces of the nasopharynx, poliovirus can multiply in specialized cells in the intestines and enter the blood stream to invade the central nervous system, where it spreads along nerve fibres. When it multiplies in the nervous system, the virus can destroy nerve cells (motor neurons) which activate skeletal muscles. These nerve cells cannot regenerate, and the affected muscles lose their function due to a lack of nervous enervation - a condition known as acute flaccid paralysis (AFP). Typically, in patients with poliomyelitis muscles of the legs are affected more often than the arm muscles. More extensive paralysis, involving the trunk and muscles of the thorax and abdomen, can result in quadriplegia. In the most severe cases (bulbar polio), poliovirus attacks the motor neurons of the brain stem - reducing breathing capacity and causing difficulty in swallowing and speaking. Without respiratory support, bulbar polio can result in death. It can strike at any age, but affects mainly children under three (over 50% of all cases).[3]

Further, the disease can have long-term impact on infected individuals, even after apparent recovery. According to the U.S. Center for disease control, "Even children who seem to fully recover can develop new muscle pain, weakness, or paralysis as adults, 15 to 40 years later."

During the first part of the 20th century, widespread epidemics of the polio virus were common:

> The first major polio epidemic in the United States occurred in Vermont in the summer of 1894, and by the 20th century thousands were affected every year. In the first decades of the 20th century, treatments were limited to quarantines and the infamous "iron lung," a metal coffin-like contraption that aided respiration. Although children, and especially infants, were among the worst affected, adults were also often afflicted, including future president Franklin D. Roosevelt, who in 1921 was stricken with polio at the age of 39 and was left partially paralyzed.[4]

By the mid-1900s, before polio vaccines were available, there were over 15,000 cases of paralysis each year in the United States alone (CDC 2023).

This dreaded disease was so feared at the time that the public was desperate for an intervention to cure and/or prevent its spread in community settings:

> In the late 1940s, polio outbreaks in the U.S. increased in frequency and size, disabling an average of more than 35,000 people each year. Parents were frightened to let their children go outside, especially in the summer when the virus seemed to peak. Travel and commerce between affected cities were sometimes restricted. Public health officials imposed quarantines on homes and towns where polio cases were diagnosed.

Although research into a potential intervention had begun many decades prior, it wasn't until the 1950s that the first vaccine was tested, developed, and then eventually distributed, leading to the virtual disappearance of the virus in the global setting. And although polio does still exist in some small, isolated areas of the globe, most of the world is now completely polio free.

The first question that will interest us here in the analysis of this example is: how do we know for certain that the polio vaccine was actually the cause of the virus' near-complete disappearance? And second, how do we know that this vaccine, or any other, is truly safe and/or effective in preventing disease? In short, we must ask the question: what is the evidence for these claims? To begin to answer this, we need to know that, in general, the way that the safety and effectiveness of nearly all medical interventions is determined is via randomized controlled trials, in which the new intervention is tested either against a placebo or against an older more established intervention. Although this is common practice for medical interventions now, the polio vaccine was the first to be tested in this way, which makes it a particularly interesting case study for our purposes. In particular, the trial of the first polio vaccine was designed in the following way:

> The statistical design used in this great experiment was singular, prompting criticism at the time and since. Eighty four test areas in 11 states used the textbook model: in a randomised, blinded design all participating children in the first three grades of school (ages 6–9) received injections of either vaccine or placebo and were observed for evidence of the disease. But 127 test areas in 33 states used an "observed control" design: participating children in the second grade (ages 7–8) received injections of vaccine; no placebo was given, and children in all three grades were then observed for the duration of the polio "season."[1]
>
> The use of the dual protocol illustrates both the power and the limitations of the randomised clinical trial to legitimate therapeutic claims. The placebo controlled trials were necessary to define the Salk vaccine—introduced by a lay organisation that has taken an activist position against the counsel of its virological advisers—as the product

of scientific medicine. The observed control trials were essential to maintaining public support for the vaccine as the product of lay faith and investment in science.

(Meldrum 1998)

Thus we can see from this historical example that even as far back as the 1950s, which was long before the advent of the EBP era, randomization and double-blinding were considered to be very important components in the design of an experimental trial, and observational evidence, which does not employ these kinds of controls, was considered to be somewhat "weaker" than RCT evidence. Yet, in this case, the designers of the polio vaccine trial opted to use both methods, experimental and observational, in their trial. The randomization in the trial was meant to satisfy the investigators that the experiment was a sound one, while the observational control component was meant to encourage the participation and trust of the public. Thus the trial designers appealed to both scientific and extra-scientific values in designing their experiment, which brings us to the question: was this experiment a well-designed one? The answer to this is somewhat complicated. In the first instance, it should be noted that it is definitely true that unconfounded, unbiased data is clearly always better than confounded, biased data, no matter its source. But the question at hand is a slightly different one – it is the question of whether or not RCT data is *more likely* to give us unconfounded, unbiased data than what we can expect from data that is gathered via observational study. By now, it should not come as a surprise to the reader that there is no perfect method of data gathering. For example, as we have already seen, randomization cannot guarantee that there will be an even distribution of confounding factors in an experiment– such a mathematical guarantee is impossible barring an infinite set. (To see why, imagine that I were to use a random number generator to distribute the undergraduate students in one of my classes into two groups. Doing so would not ensure, for instance, that everyone with red hair, or everyone over 6 feet tall, did not all end up in one group.) The same is true for RCTs – because we do not know what factors might be potentially confounding, we must rely on randomization to help to mitigate this, but we cannot rely on it to *guarantee* it, and thus not even a well-designed RCT can guarantee that confounders won't enter into the experiment.

However, when it comes to the elimination (or at least mitigation) of bias, the situation with RCTs seems to be more promising. We know that both researchers and participants (Stone et al. 2005; Holman 2015) in trials can be subject to what are known as expectation effects. As we have seen, participants who suspect that they are receiving a placebo often experience worse outcomes than those who suspect that they are receiving the actual treatment (even when they are not). These particular expectation effects are what are known as the "nocebo" and the "placebo" effects, respectively. It is also possible that:

> If clinicians involved in trials are allowed to decide to which arm of the trial a particular patient is assigned to then there is the possibility that, perhaps subconsciously, they will effect a selection that distorts the result of the trial and gives an inaccurate view of the efficacy of the treatment. They might, for example, having a view on the effectiveness of the new drug and also likely side effects, direct patients that they know to one arm or the other because of the perfectly proper desire to do their best for each individual patient, or because of the entirely questionable desire to achieve a positive result so as to further their careers or please their (often pharmaceutical company) paymasters.
>
> (Worrall 2004)

Thus, it does seem that double-blinding is a helpful way to ensure greater efficacy in trial data, by mitigating the above sorts of effects. What this means is that we need to think carefully about how our experimental designs influence the evidence we gain from them. Because there is no such thing as a perfect experiment or a perfect data set, this should lead us to be pluralists about what counts as scientific evidence. Being a pluralist about evidence means looking at all of the available forms of evidence, whether experimental, observational, theoretical, or expert opinion-based (all of which will be discussed in detail in later chapters), and being willing to fit them together (or weigh them against one another) as needed, in order to strengthen our evidence base as much as possible. However, being a pluralist about evidence is not the same thing as being a relativist regarding evidence. Generally speaking, relativism is the view that "truth and falsity, right and wrong, standards of reasoning, and procedures of justification are products of differing conventions and frameworks of assessment and that their authority is confined to the context giving rise to them" (https://plato.stanford.edu/entries/relativism/). From this, we can see that relativism is a subjective view. However, here I have argued that when it comes to science (and public policy), we should adhere to objective standards of evidence. What this means is that while we might admit more than one form of evidence into our deliberations, this does not mean that all "evidence" is on a par. It also does not mean that just because scientists sometimes disagree as to what counts as evidence in a given situation, that there is no objective way of resolving the disagreement. As an analogy, imagine that you and I disagree on the answer to a math problem. That doesn't mean that there isn't any answer! We could both be wrong, or one of us could be right. The same goes for evidential disagreements, and for ethical ones as well.

What, then, about the ethical considerations that went into the design of the polio vaccine trial? Were they ultimately helpful or harmful in the process of data acquisition? To begin, when it comes to experimental design, we should note that considerations of ethics and evidence are *always* intricately intertwined – no matter what the experiment. Further, this is the case whether the researchers designing the trial realize it or not. One specific way in which these factors intertwine is when addressing the question of how much or what kind of evidence is enough to inform a policy decision. In other words, we often need to know not only how to gather quality evidence, but also when it is time to stop gathering it. As we will see in more detail in the following section, once this level of evidence has been reached, it is often *unethical* to attempt to gather more evidence from a trial. One reason for this is that in any randomized trial, even when they are minimal, risks do exist for participants, which means that any unnecessary, or unnecessarily long-term, trial should be avoided in an effort to minimize risk for the participants involved.

In the case of the polio vaccine trial, the reason for including an observational component into the trial was primarily an ethical one, and yet it did also have important scientific ramifications as well in that it increased public support for and participation in the trial. Without an adequate sample size, the trial data would not have been as compelling. This gives us a simple and clear example of how an appeal to an ethical (non-scientific) value can actually have scientific benefit, although this is, of course, not necessarily always the case.

Evidence and values

As we have seen, the activity of gathering and evaluating scientific evidence always involves the consideration and incorporation of extra-scientific concepts. To be more precise, scientists must appeal to these sorts of concepts, values, and priorities when deciding:

1 What counts as evidence.
2 How evidence should be gathered.
3 How much scientific evidence is needed prior to making an applied policy decision.

However, coming to terms with the fact that extra-scientific concepts and values play a role in the scientific enterprise of gathering data makes many scientifically minded people feel uneasy, perhaps because they worry that these values will somehow threaten the rigor and/or objectivity of the experiment in question or otherwise diminish the quality of the evidence gathered through it. This concern is certainly understandable, and thus worth addressing here in more detail. To begin, we should recognize that just because a concept or value is not scientific, this does not mean that it is not objective. Recall that objectivity simply means "not dependent upon the subject," and moral values can (and should) have this quality too. So while it is the case that scientists are sometimes required "to make choices that are not completely settled by the available [scientific] evidence but that serve some ethical or social values over others" (Elliott 2017), this in no way should be understood as necessarily or inherently detrimental to the scientific process of gathering data.

With this in mind, let's take the above three points in turn. First, what is the role of extra-scientific concepts and values in determining what counts as scientific evidence? We have seen that strict adherents to the EBP practice paradigm often speak as if the *only* way to get quality scientific evidence is via an RCT. However, even these strict adherents themselves acknowledge that in some cases RCTs are impractical or even unethical. For example, when it is known ahead of time what the outcome of an RCT will be, then it would be unethical to run it, both because it would be a waste of money and because it could potentially put trial participants at risk for no reason. Worral (2004) gives an excellent example of such an unnecessary, and ultimately dangerous trial:

> A persistent mortality rate of more than 80% had been observed historically in neonates experiencing a condition called persistent pulmonary hypertension (PPHS). A new method of treatment – "extracorporeal membraneous oxygenation" (ECMO) – was introduced in the late 1970s, and Bartlett and colleagues at Michigan found, over a period of several years, mortality rates of less than 20% in infants treated by ECMO.
> [...]
> Despite the appeal of the treatment, and despite this very sharp increase in survival from 20% to 80% the ECMO researchers felt forced to perform an RCT despite the fact that their experience had already given them a high degree of confidence in ECMO. They felt compelled to perform a trial because their claim that ECMO was significantly efficacious in treating PPHS would, they judged, carry little weight amongst their medical colleagues unless supported by a positive outcome in such a trial.
> These researchers clearly believed that, in effect, the long established mortality rate of more than 80% on conventional treatment provided good enough controls – that babies treated earlier at their own and other centres with conventional medical treatment provided sufficiently rigorous controls; and hence that the results of more than 80% survival that they had achieved with ECMO showed that ECMO was a genuinely efficacious treatment for this dire condition. [...]

But, because historically controlled trials are generally considered to carry little or no weight compared to RCTs, these researchers felt forced to go ahead and conduct the prospective trial.

The results turned out to be disastrous:

As it turned out, the first baby in the trial was randomly assigned ECMO and survived, the second was assigned CT and died. The 1985 study reported a total of 12 patients, 11 assigned to ECMO all of whom lived and 1 assigned to CT who died. (Recall that this is against the background of a historical mortality rate for the disease of around 80%.)

As Worral goes on to argue:

Ethics and methodology are fully intertwined here. How the ethics of undertaking the trial in the first place are viewed will depend, amongst perhaps other things, on what is taken to produce scientifically significant evidence of treatment efficacy. If it is assumed that the evidence from the "historical trial" (i.e. the comparison of the results using ECMO with the earlier results using CT) was already good enough to give a high degree of confidence that ECMO was better than CT, then the ethical conclusion might seem to follow that the death of the infant assigned CT in the Bartlett study was unjustified.

But if, however, it is taken to be the case that

...the only source of reliable evidence about the usefulness of almost any sort of therapy... is that obtained from well-planned and carefully conducted randomized... clinical trials.

(Tukey (1977); emphasis supplied)

then you're likely to have a different ethical view, even perhaps that

the results [of the 1985 study] are not... convincing... Because only one patient received the standard therapy,...

(Ware and Epstein 1985)

Many commentators in fact took this latter view and concluded that

Further randomized clinical trials using concurrent controls and... randomization... will be difficult but remain necessary.

(ibid.)

We can see from this stark example that the determination of what counts as evidence – even when it comes to RCTs – always has an ethical component. When an RCT is unethical to perform (and, of course, under what conditions this is the case is itself highly controversial) then we might be forced to look for other sources of "evidence." But this itself should give us pause. What makes us certain that these other sources are less efficacious, and should therefore only be appealed to as a last resort? In cases like ECMO for PPHS and penicillin for strep throat, even setting ethical considerations aside, we can see that we really do not even need RCT data because we already have strong observational evidence that these therapies are both safe and effective. And although the evidence is observational, rather than experimental, it is both extensive and objective, in both cases. In effect, we know that these two therapies work because we have (a) watched them work and (b) know the mechanism by which they work. According to EBM, we do not need to know the mechanism by which an intervention works in order to establish its efficacy. In fact, it is a hallmark of the EBM paradigm that it is explicitly not concerned with mechanistic evidence (which is evidence about the way things work), or indeed with theoretical evidence of any kind. Instead, RCTs are designed to tell us whether or not a treatment works, not *how* it does (Kennedy and Malanowski 2018). On the EBM paradigm, then, mechanistic reasoning and other forms of theoretical evidence are considered to be far inferior to evidence gathered from randomized trials. Essentially, the EBM viewpoint is that mechanisms either don't matter or don't matter much. Instead, what matters, according to EBM, is whether or not an intervention is (a) safe and (b) effective. However, contrary to this view, it seems that given many historical examples, such as with ECMO and penicillin, observational evidence, together with mechanistic reasoning can also, in many cases, help us to establish both of these things. Thus, while the aim of establishing the safety and effectiveness of medical and other interventions is certainly the right one, there is more than one potential way to do this. And further, as the example of ECMO shows us, the choice of whether or not to run an RCT, or to instead rely on evidence derived from observation plus theoretical reasoning, will often have an ethical component: when we truly do not have sufficient evidence to answer the question of whether or not a given intervention is safe and/ or effective, then a trial should be run. But when we do already have this answer – even if the answer is derived from observation and/or mechanistic reasoning – then there is no reason to put potential trial participants at risk.

The role of values in the process of gathering evidence

As we have seen from this example, one key area in which values must be incorporated into the process of gathering scientific evidence is when human participants are a part of that process. To further elaborate on this point, we must recognize that the participants in any experimental trial must be informed about what the trial will entail for them and must be able to freely consent to participation in it. Further, potential participants need to understand that their participation in a trial is not the same thing as their receiving an intervention, which is, unfortunately, a common confusion that arises in the context of clinical research:

> A central tenet of the ethics of clinical research since the Belmont Report has been the separation of therapy from research (Emanuel et al. 2000; Emanuel 2003). [...]

> It is generally accepted that, in order for clinical research to be ethical, therapy must be detached from research, in practice and in the understanding of research subjects.
>
> (Kennedy and Cwik 2021)

However, patients often misunderstand, or in some cases are (intentionally or not) led by the recruiting researchers to believe, that their participation in a trial will provide them with the intervention being tested:

> Patients' participation in research because they mistake it for therapy is known as the *therapeutic misconception* (Appelbaum et al. 1987; Miller and Rosenstein 2003, McCormick 2018). The therapeutic misconception raises significant problems for clinical research; it may compromise informed consent, particularly in cases where participants may believe that participation in the trial is actually tantamount to a novel form of treatment, when in fact they may be assigned to a control group and may receive little to no (medical) benefit from the trial at all.
>
> (Kennedy and Cwik 2021)

Thus, researchers have an ethical duty to make sure that participants clearly understand what being enrolled in a trial means for them. But this is just one example of a way in which ethical considerations enter into the process of evidence gathering, and an appeal to extra-scientific values is not limited to the issue of informed consent. For example, even the choice of what research question to pursue, or attempt to answer, in the first place, has an inevitable value component. In particular, financial considerations are often involved in this decision-making process. For example, many nutritional researchers have recommended, on the basis of mechanistic reasoning, vitamin C to prevent or treat mild infections such as the common cold (and some practitioners actually even "prescribe" it). However, a quick search on PubMed will reveal that there "is no evidence" that supplementation with vitamin C is effective for these purposes. But what this actually means, is *not* that we have evidence that vitamin C *does not* work, but that no one has run an RCT to decide whether or not it does. The reason for this is simple: there is no money to be made from such a trial. RCTs for interventions, in general – whether medical or otherwise – are costly and someone must pay to run them, and no private corporation or government granting agency is interested in running a trial that will not ultimately be financially beneficial. New drugs are awarded patents and can generate substantial revenue once approved, while old drugs that are implemented for new purposes, or supplements that are "prescribed" for medical conditions, such as Vitamin C for the common cold, have very little, or no, revenue-generating potential. Thus trials for these sorts of interventions are rarely if ever conducted.

The Vitamin C example highlights an important ethical issue in scientific research, which is that the choice of what to research is itself an extra-scientific question. And while it is no secret (Oreskes and Conway 2010; Kitcher 2011) that monetary considerations certainly do factor in to these decisions of which questions to research, we must be careful to ensure that they are not the only considerations that do, or that they always take a primary role. Instead, as we will see in more detail in following chapters, we can and should appeal to other social and public values as well in deciding which experiments to conduct.

In addition to factoring in to the decision of what questions to research, extra-scientific values also have a role to play when scientists are deciding how and when to report data in any given situation (Elliott 2011, 2017). These decisions depend upon several factors. One is the level of risk of the situation at hand, and another is time-sensitivity. Regarding risk, imagine the situation of scientists who are concerned about climate change as well as when and how to act to prevent its worsening. In this case, while there is:

> significant uncertainty, especially with respect to the upcoming effects of anthropogenic global climate change; it is undeniable that some of these upcoming effects will have social consequences; and some of these looming social consequences are readily foreseeable.
>
> (Elliott 2017)

Because climate data has practical implications, when and how scientists report this data to the public and other stakeholders matters. Because of this, this data is covered by what is known in philosophy of science as "the argument from inductive risk." According to this argument (Rudner 1953; Douglas 2009, 2013):

> scientists must set standards of evidence for inferential decisions in science according to the possible consequences of error—including both

false positives and false negatives—at least in those cases where there are predictable social, political, or ethical consequences of such error.

That is, scientists must weigh what would happen if they were to *either* over or under estimate the risk of climate change on human populations. To do this, however, they must appeal to extra-scientific concepts and ideas – an appeal to logic alone is not enough:

> Evidence, logic, and epistemic values can tell us something about the strength of support for some claim, but they alone cannot compel a scientist to make the choice to assert, infer, accept, or endorse that claim. Thus, according to the argument from inductive risk, the scientist *qua* scientist must make classically normative judgments.
>
> (Havstad and Brown 2017)

Here again, we see how values enter into the scientific process – even into the reporting of scientific data:

> The fundamental requirements of empirical adequacy and logical consistency alone cannot compel scientists to do things one way rather than another—to choose this over that methodology, this characterization of ambiguous data over this or that conceptual framework, a higher or lower standard of evidence, etc. Rather, the scientific process involves a series of unforced choices which lead to results that, while significantly constrained by logic and evidence, are still highly contingent on the set of prior choices made. Such choices often incorporate, either directly or indirectly, non-epistemic as well as epistemic value judgments.
>
> (ibid.)

Further, this means that, in practice, there is no one set way to make such value judgments.
   Instead:

> As the last half-century or more of philosophy of science has shown, these value-laden choices are generally made on the basis of a mix of background assumptions, methodological conventions, tacit knowledge, research tradition, etc., but they are nonetheless choices, in the most basic sense, that could be made differently than they are.
>
> (ibid)

Regarding the concern of data release in a situation of time-sensitivity, this is an issue that we have seen play out since the beginning of the COVID-19 pandemic. When time is of the essence, sometimes scientific data will need to be reported sooner than might have been necessary in a non-emergent situation. The decision of when to release data, like the decision of how to release it, is also subject to the argument from inductive risk. In situations such as these, scientists will be required to weigh whether early or later data reporting will increase or reduce risk to the public and other stake-holders. What we can see then from this brief discussion of data and risk is that not even scientists can escape engagement with extra-scientific conceptual and moral reasoning! However, when they do engage in such reasoning, especially when it comes to decisions regarding when and how to report their data, they should be honest and transparent about this, so as to foster, rather than jeopardize, stakeholder and public trust. This is because, again, engaging in moral reasoning is not a problem for science, however, hiding it, in some cases, can be (as we will see in more detail in Chapter 5), because doing so can erode public trust in the scientific expertise.

Conclusion

From the discussion in this chapter we have seen that in order to most effectively answer questions of scientific evidence and public action, we should aim to be pluralists (but not relativists) about what counts as evidence in a scientific context, and also be willing to openly and honestly appeal to non-scientific, ethical values, when this is required. This means, first of all, that the more, and the more types of, evidence we can draw upon in order to inform our policy decisions, the better. In practice, then, we should not limit our view of what counts as scientific evidence to experimental or RCT evidence only, and instead should recognize that observational evidence, theory, and even expert opinion (as we will see in Chapter 5), in many contexts, is worth appealing to in order to best inform our policy decisions. Thus the process of learning how to identify, incorporate, and weigh various kinds of evidence for scientific claims is of pivotal importance especially when the scientific data in question will ultimately be applied to public policy.

Further, when it comes to the necessary incorporation of philosophical values such as ethical and social ones into the process of gathering scientific evidence, as is outlined in the argument from inductive risk, we should not be afraid that doing so will somehow "mess up" or contaminate the acquisition of objective data. While an appeal to extra-scientific values will *always* be required when applying science to policy, this does not mean that the process is thereby without rigor or is in some way compromising of objective inquiry. On the contrary, philosophy is an endeavor that aims for both rigor and objectivity in just the way that science does. And thus incorporating extra-scientific concepts, such as philosophical and/or ethical ones into the process of developing science-informed public policies should not be seen as a detriment, but instead as a beneficial and important component of the enterprise, one that serves to give us better, stronger evidence than we would otherwise have, and which we can then apply to our public policy decisions.

Notes

1 https://public.wsu.edu/~taflinge/evidence.html.
2 The objective/subjective distinction is different from the descriptive/normative distinction which we will see in more detail later.
3 https://www.who.int/teams/health-product-policy-and-standards/standards-and-specifications/vaccines-quality/poliomyelitis
4 Centers for Disease Control and Prevention. "What is Polio?" https://www.cdc.gov/polio/what-is-polio/index.htm

3   Interpreting the science

Because raw data as such have little meaning, a major practice of scientists is to interpret data [...] Such analysis can bring out the meaning of data—and their relevance—so that they may be used as evidence.[1]

Introduction

In the previous chapter, we discussed in detail the question of how to identify what counts as scientific evidence. We also saw in that chapter some of the ways in which scientific evidence can be used to support claims about scientific and/or moral matters. In this chapter, we will begin by taking a step backward, so to speak, by taking a look at the process of interpreting scientific data, which takes place before that data can be used as evidence, as well as at the overarching question of why this interpretation of data *matters*, particularly within the context of its application to public policy decisions.

To begin, it is important to note that *all scientific information requires interpretation* – no scientific data interprets itself: in other words, science can tell us what is the case, but it cannot, on its own, tell us what its data outputs mean (nor what, if anything, to do with that data). Further, this point holds equally for data that is acquired from scientific models, from observational studies, and from experiments. To put it another way, as we have already seen, scientific data is often used as evidence, for various purposes and applications. However, there is a gap between the data that science provides us with and evidence, particularly when it comes to application of that data/evidence. In other words:

Evidence does not reside only in the world where science is produced; it emerges in the political world of policy making, where it is interpreted, made sense of and is used, perhaps persuasively, in policy arguments.[2]

DOI: 10.4324/9781003311072-3

Essentially what this means in practice is that scientific evidence is derived, not from data alone, but rather from data *plus* certain extra-scientific considerations. Furthermore, as we will see in the example that follows, these extra-scientific considerations enter into the scientific process of discovery even before data is acquired, for instance, in the choice of what question or questions to research in the first place (we might think of this as "pre-interpretation"). When deciding what to research, scientists are making a "non-scientific" decision, so to speak: they might decide to address a certain question because they believe that it will have public health ramifications, or because it is a financially lucrative to pursue, or because it is potentially Nobel prize worthy, etc. But whatever the reason for choosing a particular area of study, the reason itself is never a purely scientific one: as we saw in Chapters 1 and 2, science provides us with a method for inquiry, but it cannot tell us what to study, or when or why.

Next, once a research question has been chosen, scientists will then decide on whether to pursue it via a process of modeling, experimentation, observation, or some combination of these things. And as we will also see in what follows, extra-scientific considerations enter during this part of the scientific process as well. And finally, once enough data has been gathered from the various methods employed, extra-scientific considerations again enter into the process. When the data has been gathered, it must then be interpreted so that we know what it *means*, and so that we can use it – as, for instance, evidence in support of a proposed science-informed policy. Before we dive into the details of an actual example intended to illustrate this process, it is first helpful to address an important question, which is the question of why we should care about whether or not scientists incorporate extra-scientific considerations into their research processes. The answer to this question is that we should care about this because if scientific inquiry as well as data interpretation requires more than "just science," then this, in turn, puts pressure on the dearly held (by many) view that science can provide us with purely objective, "value-free" facts. In other words, it forces all of us – scientists and non-scientists alike, to acknowledge that the method of science is an inherently human and value-laden endeavor. However, at the same time, we should also realize that this is as it *should* be: when the evidence that is derived from the data we gather is subsequently used to implement or influence public policies, not only will it always involve a normative component, but this is a good thing! We will see in more detail why this is the case in Chapters 4 and 5, but suffice it to say for now that because such policies are implemented by and for people, in order to bring about certain states of affairs that we deem to be beneficial, this process is an inherently normative endeavor. In the example that follows, we will see this point more clearly by taking a closer look at this process of project choice, data interpretation, and subsequent application, while paying particular attention to the various points and ways in which extra-scientific considerations enter into this process.

Example: secondhand smoke and bans on smoking in public places

It has been well known since at least the time of the 1964 Surgeon General's Report[3] that secondhand smoke, which is defined by the CDC as "the combination of smoke from the burning end of a cigarette and the smoke breathed out by smokers," has serious and detrimental health effects when it is inhaled. Secondhand smoke contains more than 7,000 chemicals and has been shown to cause cancer, heart disease, stroke, and premature death among nonsmokers (USDHHS 2006). Further, it has also been shown that there is no risk-free level of exposure to secondhand smoke – inhaling any amount is considered to be unsafe, although, of course, inhaling more is considered to be worse in terms of health effects (USDHSS 2006). Furthermore, the effects of secondhand smoke have been found to be particularly dangerous for children, and studies have shown that children whose parents smoke get sick more often, and are more likely to have asthma or to die of SIDS (American Academy of Pediatrics 2005; Huang and Chen 2006; Moon 2017) than are children who have parents that do not smoke. And finally, all of this seems to be particularly bad when we take into consideration the fact that those who suffer these detrimental effects generally are not at fault – because they often do not have any viable way to avoid it. Thus, we can begin to see then that the "problem" of secondhand smoke inhalation clearly has both scientific and ethical components. The scientific component of the problem is that the scientific data that we have suggests that chemicals from inhaled secondhand smoke cause effects in humans. The ethical component of the problem is that we deem these effects to be bad, that is, detrimental, or unhealthy, because we believe that they are harmful to both the individuals that experience them and to the communities in which they reside, and, further, because they affect people who, for the most part, are not in any way responsible for, or able to prevent, these ill effects.

We should note here that the identification of this type of two-pronged scientific/normative problem is always the first step in any potential application of science to public policy, because if there is no ethical component to the problem, then no policy intervention is necessary – purely scientific issues are addressed in other ways within the scientific community. However, once this sort of dual problem is identified, then as a first step toward an attempt to solve it, scientists will often build and then subsequently evaluate a model – whether formal or informal – of potential solutions and/or policy interventions designed to address it. Scientific models are useful in this endeavor because they allow us to make predictions about what *would be* the case if certain variables (such as differing types of policy interventions) were to be changed or implemented, and this, in turn, allows us to make predictions about what would happen if we were to implement them in the part of the world that the model is intended to represent. Scientific models allow us do this by giving us a way to combine as background information both previously established scientific data as well as various generalizations and/or idealizations, intended to make the model simpler to work with, and then, against this background, we can superimpose various proposed interventions – to see what would happen if they were to be implemented. In other words, if we decide that our scientific data on a given issue points to something that is harmful or problematic, we can then create a model to facilitate predictions and/or explanations of what would happen if certain policy recommendations were to be put into effect.

For example, in the case of secondhand smoke, scientists first decided to pursue an informal model of potential policy solutions in the early 1970s, because they were concerned about the negative health effects of inhaling the smoke. The data that they had gathered predicted that, if public smoking bans were to be widely implemented, then these bans would significantly decrease adverse health outcomes in the public by decreasing secondhand smoke inhalation (CDC 2007; Haw and Gruer 2007). However:

> At this time, the scientific evidence on the health effects of exposure to secondhand smoke was limited. Studies starting in the late 1960s had shown adverse effects of maternal smoking on the developing fetus and on children exposed to secondhand smoke in smoking households. However, it was not until the following decade that a critical mass of scientific evidence emerged linking exposure to secondhand smoke with cancer and other chronic health effects among nonsmoking adults. In 1980 and 1981, scientific journals published epidemiologic research from Greece, Japan, and the United States finding that those who breathed "environmental tobacco smoke" suffered from decreased lung function and increased risk of lung cancer.[4]

Nevertheless, even in the 1970s scientists were able to use what data was available in order to predict the ways in which smoking bans would potentially affect public health outcomes. This shows us how, even with incomplete data, scientific modeling and prediction can be useful in helping to inform public policy interventions. And indeed, because of these scientists' early modeled predictions, and their subsequent recommendations, the first restrictions on smoking in public places, and government buildings, were implemented in the mid-1970s (IOM 2010). For example, in 1975, the state of Minnesota enacted the Minnesota Clean Indoor Air Act, making it the first state to restrict smoking in most public spaces. Many states soon followed suit with similar restrictions:

> In 1973, the Civil Aeronautics Board ordered domestic airlines to provide separate seating for smokers and nonsmokers. In 1974, the Interstate Commerce Commission ruled that smoking be restricted to the rear 20% of seats in interstate buses. States and localities also began to impose restrictions. In 1973, Arizona became the first state to restrict smoking in some public spaces. In 1974, Connecticut enacted the first statute to restrict smoking in restaurants. In 1975, Minnesota passed a comprehensive statewide law "to protect the public health, comfort, and the environment by prohibiting smoking in public spaces and at public meetings except in designated smoking areas." In 1977, Berkeley, Calif, became the first local community to limit smoking in restaurants and other public settings.[5]

Further, these bans became progressively expansive, and stricter, over time, even as they increasingly gained public support:

> Once these efforts gained momentum, new legislation spread rapidly. The recognition of exposure to secondhand smoke as a health risk to nonsmokers meant that the issue was no longer merely one of individual choice. People responded differently to risks that were imposed on them involuntarily. The existence of victims of cigarette smoking fundamentally altered the discussion about the right to smoke, and state and legal intervention was seen as entirely appropriate. There was also substantial public support for enacting restrictions on smoking in public spaces. As early as 1970 (before any Surgeon General had spoken out about harm to nonsmokers), 58% of men who had never smoked and 72% of women who had never smoked responded 'strongly agree' or 'agree' that smoking should be allowed in fewer public spaces than it was at the time.[6]

This momentum carried through for the next several decades. For instance, in 2003 the state of New York banned smoking in all indoor public places, including restaurants and bars, which had been exempted in previous bans, and by 2018, 29 out of 50 states, as well as the District of Columbia, had implemented some form of public smoking ban.

We should notice here, however, that even though scientific data that was available as early as the 1970s associated certain negative health outcomes with secondhand smoke inhalation, and indeed eventually led to the widespread implementation of public smoking bans, this data required interpretation – via the incorporation of extra-scientific considerations – prior to its application. In other words, the scientific data on secondhand smoke inhalation provided us with a description of what was the case (prior to any attempted intervention) and also allowed us to make predictions about what *could be* the case, if certain actions were to be taken to mitigate it. However, the data did not, and indeed could not, on its own, tell us whether or not these actual or predicted states of affairs were good or bad to begin with, or whether or what, if anything, should be done to change that. And this is because judging a state of affairs to be good, bad, or problematic, as we have seen, always has an ethical component and thus is an endeavor that requires extra-scientific considerations – science alone is not able to make this judgment. Thus, in this specific example we can already see illustrated the more general point that extra-scientific considerations enter into the process of applying data to policy in many places. First, they enter into the choice of what question to study (which in this case was the question of what were the health effects of secondhand smoke exposure), and they also enter into the choice of whether or not to do something about negative effects, once they are identified.

More specifically, in the case of secondhand smoke, scientists chose to study this question because of its potential public health ramifications. That is, they deemed it to be not only a scientific problem, but an ethical one as well. Second, they decided to pursue potential policy interventions to solve this problem by building an informal model that would help them to predict the impact on individual and community health outcomes that differing policy interventions would have. Not surprisingly, the model predicted that less exposure to secondhand smoke would in fact lead to better overall community health outcomes. Further, the scientists who built the model reasoned that if we did nothing to curb the current levels of second-hand smoke inhalation then the current rate of illness and death caused by it would remain unacceptably high, which was another normative (non-scientific) judgment they made. And then because of this, they proposed various potential policy interventions: that is, the scientists hypothesized that banning smoking in public places would reduce the negative smoke-related health implications amongst non-smokers, and so they proposed that bans should be implemented, as they eventually were.

Next, after these bans were implemented, it became incumbent upon scientists to determine whether or not they were effective. In other words, they needed to ask the question of whether or not the bans were successful in improving public health outcomes, as predicted. Initial data overwhelmingly showed that they were (Mackay et al. 2010; Hahn 2010). However, even in spite of this positively correlated data, which seemed to show that public smoking bans were indeed accomplishing their intended aim, many were left asking the question of whether or not we *really* could say for certain that these bans were the *cause* of the observed improvement in certain public health outcomes, and thus worth continuing. Here it should be noted that this sort of question is not in any way specific to this particular example. Instead, it is one that nearly always arises when evaluating science-informed public policies precisely because cause and effect is notoriously difficult to establish in any instance, or area, of scientific inquiry. Because of this, in general, in an effort to answer the question of whether or not a given policy intervention is effective, after it is implemented we will want to seek out at least some level of confirmation that it is, on the balance, more helpful than harmful, in order to decide whether or not it should be continued, even if or when direct causation cannot be established. In the case of public smoking bans, in particular, later research (such as Thomson 2006) did, in fact, show that there were fewer smoke-related negative health outcomes after the public smoking bans had been in place for several years, suggesting that they were beneficial. However, we also know that secondhand smoke exposure in the population was decreasing anyway,[7] likely because of a combination of decreased rates of cigarette smoking, and increased awareness of the risks of secondhand smoke exposure, in addition to the adoption of smoke-free laws prohibiting smoking in workplaces and public places in many states and localities, so this data did not prove causation. To be more specific, this is because:

> the health of nonsmokers after the implementation of a smoke-free policy can be affected not only by reduced secondhand-smoke exposure but also by concurrent changes (such as home smoking bans and decreases in smoking by people in other environments) attributable to increased awareness in the community, increased spontaneous cessation, and higher cessation success rates. The latter factors might have additional implications for the period over which follow up is performed because their own timing might influence the effectiveness of a ban. Therefore, in evaluating and interpreting studies of the effects of smoking bans on health outcomes, the other concurrent activities must also be taken into consideration. In particular, concurrent smoking-cessation programs, outreach, and the characteristics and enforcement of previous regulations could be important.[8]

However, even though follow-up studies after smoking bans were implemented did not definitively or directly establish a cause and effect relationship in this particular case, we do know that public smoking bans play at least some role in reducing secondhand smoke *exposure*, even when we take into account other factors, such as the ones mentioned above. To summarize, the reason that public smoking bans were implemented in the first place is because it was thought, given the predictions of informal models, that they would improve public health outcomes. And, while it is difficult to say whether or not this improvement was due *only* to the bans, the data does in fact strongly suggest this. For example, in 2003, a statewide comprehensive smoke-free law prohibiting smoking in all indoor areas of workplaces, restaurants, and bars was implemented in New York (this was well after smoking rates had begun to decline for other reasons). A subsequent study found that salivary cotinine levels among nonsmoking adults decreased by 47.4% within one year after the law took effect.[9] and another study found that salivary cotinine levels among nonsmoking adult workers in the area decreased by 85% within one year after the law took effect (ibid.). Thus, while there very well could be other reasons for the reduction in secondhand smoke inhalation, smoking bans clearly do play a significant role. Given this, we can reasonably assume that smoking bans decreased the rate/level of smoke inhalation. From this fact, it further seems reasonable to infer that less smoke exposure, in turn, leads to improved health. Thus, this is a form of indirect evidence that the bans were and are playing a role in the observed community health improvements, even in the absence of directly establishing cause and effect, which is something that we will nearly always fall short of in any scientific inquiry.

This then raises the larger question of how we should proceed when it comes to the implementation of public policies that, given our best available evidence, we believe to be beneficial, even if we cannot establish with absolute certainty that they are. In this case, the specific question was whether or not the bans on smoking in public places should be continued: if they were effective, and, on the balance not harmful, then it would make sense to continue them. Whereas, on the other hand, if they were superfluous, or in some way harmful or otherwise detrimental to the public, then it might not.

In some situations of epistemic uncertainty, we might want to conduct further research studies in order to gather enough data to answer the question of whether or not to continue the implementation of a particular public policy. However, in other cases, such as this one, it is often more appropriate, and timely, to appeal to community values in the decision-making process instead:

> Lack of certainty appears to be a common feature of many areas of science at the interface with society. For example, economists, climate scientists and environmental researchers are often asked to take a stand on contentious socially-relevant questions. The demands of the situations where scientific expertise is needed may not allow waiting for additional evidence – or the means for acquiring further data might not be available. Yet expert opinion is called for. Advice given in these situations should be evaluated according to criteria that take into consideration the need of acting under uncertainty and the non-epistemic consequences of decisions.
>
> (Jukola 2022)

Further, the details of what criteria are appealed to in any given case will depend on the specific situation. For instance, in this case there was, and still is, strong public support for bans on smoking in public places: a Gallup poll conducted in 2011 found that 59% of American adults support these bans and, in addition to this, nearly 25% support the total illegalization of tobacco smoking. Further, these numbers have remained fairly consistent since that time.[5] And this is something that it seems reasonable to appeal to in the decision-making process regarding whether or not to continue the bans. There is also available research data that shows that bans on smoking in public places do not hurt the hospitality industry as some had initially feared they might (Hahn 2010; Marti and Schlapfer 2014). Thus, there is, in this case, strong public support for policy that, given the available data, seems to be, on the balance, beneficial rather than harmful.

However, not everyone was, or is, in favor of public smoking bans. Even as early as 1975, Gary Huber (1975), who in later years "would emerge as a sharp critic of the developing public health consensus on the risks of tobacco smoke to nonsmokers, concluded that beyond the "psychogenic" effect of exposure to [secondhand smoke], the questions centering on the potential biological effect "remain unanswered"" and thus we did not have the needed scientific support to back up public smoking legislation. In addition to this, writing five years later, when a study was published that found that secondhand smoke impaired respiratory function, Claude L'Enfant and Barbara Liu of the National Heart, Lung, and Blood Institute asserted that until that time, "[t]he case against smoking in the environment has often been anecdotal, based on annoyances, feelings and sometimes more objective physical reactions such as eye and nose irritation." And, in fact, even though there was solid evidence that maternal smoking had damaging effects on fetuses, and that children could be harmed by secondhand smoke in the home, L'Enfant and Liu stated that, "Generally speaking, the evidence that passive smoking in a general environment has health effects remains sparse, incomplete, and sometimes unconvincing" (Grier 2017).

A few years later, in the 1980s, the then surgeon general C. Everett Koop weighed in on this debate, stating that he "was mindful of the controversy that surrounded the issue of passive smoking—a controversy fueled by the tobacco industry." It was Koop's view at that time that these detractors were politically and financially motivated to "interpret" the scientific data in a way that minimized the positive effects of public smoking bans, and that therefore they should be ignored. His view was that,

> for the purposes of the public health, the data were good enough, and the costs of inaction were too great: "Critics often express that more research is required, that certain studies are flawed, or that we

should delay action until more conclusive proof is produced," yet, "As both a physician and a public health official it is my judgment that the time for delay is past, measures to protect the public health are required now."[10]

The controversy over secondhand smoke legislation continues even today with current detractors continuing to argue that these bans were never truly supported by good science, while citing more recent studies that seem to show that smoking bans do not have (at least some of) the positive health effects on individuals that proponents of the bans claim that they do. For example, some have pointed out that a 2008 study in New Zealand[11] "found no significant effects on heart attacks or unstable angina in the year following implementation of a smoking ban; hospitalizations for the former had actually increased," and a 2013 article in the *Journal of the National Cancer Institute* declared that there is no "clear link between passive smoking and lung cancer" (Peres 2013). Even these detractors, however, do not argue about the data regarding childhood asthma and SIDS, for instance, which seems to be accepted by everyone, on both sides of the debate.

One of the important things that this example of the controversy over public smoking bans highlights for us more clearly is that ambiguity or uncertainty in the available scientific data on any given issue is often used, or even exploited by, those who are "ideologically opposed to regulatory interventions" more generally (Bayer and Colgrove 2002). This is particularly likely to occur in the context of a liberal democracy, such as the United States, where policies that are perceived as overtly paternalistic are likely to be rejected in the public sphere. Consider this quote from a 2017 article in *Slate Magazine*:

Early arguments for smoking bans at least paid lip service to the idea that restrictions were necessary to protect unwilling bystanders' health. But as bans have grown ever more intrusive even as the case for expanding them has withered, that justification has been revealed as a polite fiction by which nonsmokers shunted smokers to the fringes of society. It was never just about saving lives.

[...]

There were good reasons from the beginning to doubt that smoking bans could really deliver the promised results, but anti-smoking advocacy groups eagerly embraced alarmism to shape public perception. Today's tobacco control movement is guided by ideology as much as it is by science, prone to hyping politically convenient studies regardless of their merit and ostracizing detractors.

[...]

It may be neither feasible nor desirable to set back the clock and permit smoking everywhere, but laws in a liberal society can accommodate the rights and preferences of smokers and business owners far better than they do now.[12]

The quote from this article illustrates for us the fact that, very often *both sides* in any debate over science and public policy will try to show that the science is on *their side*. However, as we have already seen in Chapter 2, scientific data alone cannot ever resolve a debate like this. This is due to the fact that science cannot tell us (a) whether or not any given situation is good or bad or (b) whether or not we should act on it, or how.

The question that then faces us is: how should we proceed under such conditions of scientific uncertainty, when it comes to public policy interventions, given that we know that people (and even scientists) will not always agree on what the data means? The short answer to this question is that we must appeal to extra-scientific values in this decision-making process. Of course, people may not always agree on these values either, however, this is always the case in a democracy, and these public values must be decided upon publicly. For example, in the case of smoking bans, it does seem reasonable to appeal to voter desires, strong public support, and the lack of harm in order to help inform the decision of whether or not to continue the policies that are in place (i.e., to decide whether or not the policies are good ones), even though the available scientific data regarding them is somewhat inconclusive. In other words, in this case, even though we do not have absolute proof of cause and effect (and again, it should be noted that this is not something we ever have in *any* case) regarding the role of these bans in facilitating better health outcomes in individuals and their communities, we can appeal to the fact that there was (and is) strong public support for these policies, that there is good evidence that they are very likely beneficial, as well as to the fact that they do not seem to have any detrimental effects, when deciding whether or not to continue to implement them, or whether or not to regard them as "good" or helpful.

This general method of supplementing scientific data with extra-scientific considerations during our decision-making process then bears on the commonly asked question of how we should weigh science vs. extra scientific values as sources of evidence when it comes to making and supporting these important policy decisions. The answer, in the first instance, is that we must evaluate each situation on a case by case basis – this is inescapable, as the details of each case, and each community, often vary considerably. Further, we must also acknowledge that we will never be in a situation in which there will be a policy decision in which extra-scientific values do *not* play a role, even though there are certainly times when such values or considerations will play more of a role than at others. In addition, some extra-scientific values might outweigh others in our considerations, depending on the ends that we aim to achieve, the issues that are at stake, and even the amount of time we have to address these issues, because some issues are urgent whereas others are not, or are less so. For example, in urgent, high-risk situations we might opt for less data collection (in the interest of saving time prior to implementing policy action) whereas when time is less of an immediate issue, maximum data acquisition might rightfully become more of a priority in our decision-making process.

Interpreting experimental data

Although in the previous example we focused on data derived from a scientific model, we should also remember that experimental data requires interpretation as well. This is the case for several reasons. First of all, as with scientific modeling, before any experiment is conducted, someone (or some group of people) has to decide that it is worth conducting, and the incorporation of extra-scientific values is an inevitable part of this decision-making process. As we have already seen, this is especially evident in the case of randomized controlled trials for medical treatments. For instance, testing medical treatments that have the potential to be financially lucrative is nearly always preferred by researchers over testing those that are expected not to be – regardless of the projected effectiveness of the treatment in question. For example, in developed countries, such as the United States and the United Kingdom, treatments for heart disease and type 2 diabetes are constantly undergoing new development and testing, whearas, on the other hand, new antibiotics rarely make it into the research and development pipeline, even though they are desperately needed to treat bacterial infections. Further, this has nothing to do with concerns regarding potential effectiveness, but is rather due to the fact that pharmaceutical interventions have the potential for enormous amounts of revenue generation when it comes to treating heart disease and diabetes, due to the very high level of demand, while this is not the case when it comes to antibiotics. In illustration of this point, what one drug company executive said about this issue is that:

> This is the one therapeutic area where when we design a brand new antibiotic, what will happen is the world will say, 'Thank you very much, that's a great contribution to human medicine, but I'm going to put it on the shelf and only use it when I really have to use it'.[13]

This means that new antibiotics are not considered to be financially lucrative, since their sale, by medical necessity, will be limited. To be clear, although they are common, financial reasons are not the only ones that factor into the choice of what experiments to conduct, in other cases, other considerations factor in as well.

Observational studies

But what about observational studies? Does the data we gather from them require interpretation as well? The short answer is yes. And the reason for this is because human researchers are subject, in general, to many different types of systematic cognitive biases – that is, to recurring errors, both logical and probabilistic, in thinking and reasoning, and biases such as these are more likely to be problematic when they are not controlled for, as in observational studies.

To see this more clearly in context, we can begin by thinking of an observational study as a kind of trial, but without the controls of blinding or randomization applied (this is the kind of study that was used to gather data in the secondhand smoke example). However, it is precisely due to the absence of these controls that this type of study is thought to be more prone to interference from biases (as well as confounding factors) than are RCTs. In other words, the concern is that, without these controls, certain biases are especially likely to be present, and to be potentially problematic in confounding the data. And yet, in spite of this, it is interesting to note that there are many who hold that observational studies, particularly in the form of case studies, can still be valuable sources of evidence (Kiene et al. 2013). In fact, some go so far as to argue that, at least in some instances, case studies can be *better* sources of evidence than even RCTs. For example, Rachel Ankeny has argued that case studies can be helpful in establishing novel medical diagnoses, "despite having a somewhat dubious reputation as a form of anecdotal evidence" (2014). This is because:

> cases serve as 'vehicles' for gathering facts and putting them into contact with each other. Think of a train caravan composed of a series of different types of cars joined together, but where it is not yet clear which cars actually will make the trip, in which order, or what in fact is the 'engine'. [...] the process of sorting through the details contained within a case helps clinicians and researchers refine hypotheses as to the underlying cause(s) of a particular condition.
>
> (Ankeny 2014)

In other words:

> when cases are brought together and compared, they often generate testable causal predictions, as similarities and differences are articulated, and information added, until the cause can be identified.
>
> (ibid)

This, in turn, means that:

> At a minimum, cases provide a way forward for clinical practitioners who need working casual attributions with some manipulable facts in order to treat subsequent patients who might present with similar patterns of symptoms until better evidence can be produced (if ever), but they also can generate clear evidence about causal hypotheses worth testing by researchers.
>
> (ibid)

The basic idea behind this argument is that case studies can, at the least, provide us with evidence "in areas where RCTs are not ethically or practically possible," (1008) as well as supplement RCT data in instances where it is available, and, at maximum, they can, in some instances, provide us with even stronger evidence than RCTs. For example, most people would consider it unethical to run an RCT to determine whether or not the inhalation of secondhand smoke is detrimental to certain health outcomes. And this is because the observational data that we have shows that it is. Thus, as with the ECMO example that we saw in detail in the last chapter, the consensus in the scientific community is that it would be unethical to run such a trial. Further, for many rare diseases, such as Addison's disease (an uncommon endocrine disorder), or common variable immune deficiency, RCTs are rarely, if ever, conducted, largely because it would not be profitable enough to run the experiment, given the small number of people who suffer from these conditions. Thus, observational and case studies in instances such as these can play a vital role in providing much needed evidence when RCTs are not being run.

But if data from case studies are particularly prone to human biases, how do we know that we can trust the data that we derive from them? One reason is because we know that biases in reasoning, whether scientific or otherwise, can be overcome via a technique known as "debiasing." To understand what this process is, and how it works, we first need to have a better understanding of what biases are in the first place. For our purposes, we can define a bias as a systematic deviation from an objective standard (Baron 2012). First, it is important to note that not all logical or judgment errors are systematic, so not all of these errors would count as biases. In other words, a mistake is not the same thing as a bias: making a few mistakes on a math test, for example, would not necessarily count as a bias, unless, of course, the mistakes were systematic ones.

To further expand on the concept, biases can take several forms, the two most common being either a logical inconsistency, consisting of two or more beliefs that contradict each other, or a misperception of reality, or of the way that things are in the actual world. A (trivial) example of a logical inconsistency would be simultaneously holding the two beliefs that I am both standing and I am not standing right now, which is, of course, logically impossible. An example of a misperception of reality, on the other hand, would be the belief that all squirrels are brown, which is clearly a misperception of the way the world actually is, since black squirrels, for instance, have regularly been observed. If mistakes of either of these kinds are applied *systematically,* then they become what are known as biases, with the potential to negatively affect research outcomes. Some examples of well-known biases in scientific research include select/allocation biases, which is when researchers either consciously or unconsciously enroll subjects in one group vs. another in a trial, expectation bias, or the expectation that the intervention that is being tested either will or will not work, overconfidence bias (Crosskerry and Norman 2008; Cassam 2017), which is an assumption that can lead to disinterest in investigating supplemental decision support in the decision-making process of how and when to apply data, and premature closure, which is the narrowing of options too early in the process of decision-making so that the most apt option is never seriously considered (Berner and Graber 2008). Further, some version of what is known as "narrow thinking," or an inability to look at the "big picture," often underlies cognitive biases (Soll et al. 2016). And this means that biases are not just simple or isolated inconsistencies or misperceptions, but are rather specific types of these things, applied in a systematic way.

Now that we have more of an understanding of what a bias is, we can begin to better understand the process of accounting for, and attempting to eliminate (at least some of), the potential influence of these systematic errors. Doing so, as we already noted, is a process which is known as "de-biasing," and it involves training researchers in the formation of new research habits (Fisk et al. 2015; Daniel et al. 2017). There are several ways in which this can be done. For example, if the bias in question (whether it is a logical one or one of misperception) is thought to be due to a faulty intuition, then it has been shown that in order to account for and correct this, researchers need to be taught to pay attention to and reflect on the situation at hand, at least briefly, before arriving at a final judgment (Kahneman and Frederick 2005). However, this is not the only thing that is needed in order to overcome a potential bias. To do this, one also needs to have at least a rough understanding of the correct rule (logical or otherwise) that applies to the situation, and thus de-biasing often requires teaching practitioners these rules (Larrick and Soll 2012). Once these things have been done, it is thought that the most effective type of education is then to follow up with domain-specific training on a decision task that will be engaged in very soon and possibly repeatedly (Soll et al. 2016).

The field of de-bias training is one that is constantly undergoing development, and at this point no one would venture to say that its methods are fool-proof. However, even its limited successes show us that the situation is not hopeless: "Bias in judgment and decision making is a common but not insurmountable human problem" (Soll et al. 2016). Thus, while all humans, and especially bright, well-educated ones (such as scientists!) are inevitably prone to a variety of cognitive biases, it is possible for these to be addressed and at least mitigated, if not completely overcome, by targeted education and training. What this requires, however, is that these biases first be recognized, so that they can then be addressed. In the context of application to public policy, what this means is that we need to recognize and account for potential biases in the acquisition of any observational data that we incorporate into the decision-making process, in order to ensure that the best science-informed polices are ultimately implemented.

To return to our example of secondhand smoke, we can now see that extra-scientific considerations enter into the process of scientific inquiry, first, in the decision of what to study, then in the decisions regarding whether or not to pursue potential policy interventions as well as in the process of interpreting the data, including in the process of post research de-biasing. And again, as we have seen, scientific data, as long as it is accurate, can tell us what *is* the case, but it cannot tell us what *ought* to be the case or what, if anything, we should do to facilitate bringing about that state of affairs. What this means, in the context of our secondhand smoke example, is that whether the data we have regarding a specific research question comes from a model, an RCT, or an observational study, it will not, on its own, be able to tell us whether or not the outcomes that it predicts are good, bad or neutral, or how we could or should try to change them: this is something that requires extra-scientific, ethical analysis. Thus, when it comes to the application of experimental data to public policy, these kinds of considerations matter because with public policy interventions our goal is always to improve actual situations for actual people. This, in turn, means that, as we will see in greater detail in Chapter 5, we need to make sure that our extra-scientific considerations foster that aim, and that, requires taking into account both individually held as well as community values.

Conclusion

In this chapter, via the examination of a specific example, we have seen that all scientific data, whether it is derived from observational studies, modeling, or other methods of experimentation, requires interpretation, in many cases both analytical and ethical, before it can be effectively applied to policy decisions and/or interventions. Further, we have also seen that the decision of what to study in the first place, when it comes to scientific inquiry, also requires us to take into account extra-scientific considerations, a process that we might think of as "pre-interpretation." More importantly, the reason that we should care about this is because it forces us to acknowledge that the method of scientific inquiry, from the choice of what to study, to data acquisition, all the way through to interpretation and de-biasing, is not ever "purely" scientific or value-free, and indeed should not be. Instead, as we have seen, extra-scientific considerations inform this process from start to finish. To be more specific, this is true regarding scientific models because these epistemic tools always incorporate idealizations or other simplifications, and because of this, their data outputs require interpretation prior to application in real-world situations. Further, scientific experiments always require us to make decisions about what to include and what to control for, and even whether or not to run the experiment in the first place, and this means that the data we derive from them also requires interpretation prior to application. Finally, even "unadulterated" observational studies are not immune to this interpretation requirement – and this is because observational studies are especially prone to influence from human biases (since they are, by definition, not "controlled") and that must be accounted for and/or "de-biased" retroactively. In sum, what this means is that in practice *all scientific information requires interpretation via extra-scientific input*. Thus, if we are to understand the scientific information we have, and eventually be able to use it as evidence in making policy decisions, we must first learn how to "convert" knowledge about the model, experiment, or observation in question to knowledge about the target system or population in the actual world that it is meant to represent, and an appeal to extra-scientific considerations is required in order to do this.

Notes

1 https://ngss.nsta.org/Practices.aspx?id=4.
2 https://nap.nationalacademies.org/read/13460/chapter/2di.
3 Although even "Before the 1964 report was released, there had been several previous statements from the Surgeon General, and several consensus statements from groups of public health scientists, all identifying cigarette smoking as a cause of lung cancer." https://www.ncbi.nlm.nih.gov/books/NBK179276/pdf/Bookshelf_NBK179276.pdf.
4 https://www.cdc.gov/tobacco/data_statistics/fact_sheets/secondhand_smoke/general_facts/index.htm
5 https://www.ncbi.nlm.nih.gov/pmc/articles/PMC1447493/
6 https://www.ncbi.nlm.nih.gov/pmc/articles/PMC1447493/

7 Measurements of blood serum cotinine (as a measure of secondhand exposure among people who do not smoke) show that exposure to secondhand smoke steadily decreased in the United States between 1988 and 2014:

- During 1988–1991, almost 90 of every 100 (87.9%) people who did not smoke had measurable levels of cotinine.[5]
- During 2007–2008, about 40 of every 100 (40.1%) people who did not smoke had measurable levels of cotinine.[5]
- During 2011–2014, about 25 of every 100 (25.2%) people who did not smoke had measurable levels of cotinine.[5]

https://www.cdc.gov/tobacco/data_statistics/fact_sheets/secondhand_smoke/general_facts/index.htm.

8 https://news.gallup.com/poll/237767/one-four-americans-support-total-smoking-ban.aspx.
9 https://www.ncbi.nlm.nih.gov/books/NBK219563/
10 https://ajph.aphapublications.org/doi/10.2105/AJPH.92.6.949
11 https://www.health.govt.nz/system/files/documents/publications/smokefree-evaluation-report-with-appendices-dec06.pdf.
12 https://slate.com/technology/2017/02/secondhand-smoke-isnt-as-bad-as-we-thought.html
13 David Payne, a vice president overseeing antibacterial drug discovery at Glaxo-SmithKline.

4   Communicating the science

All science exists in a social context.

(Oreskes and Conway 2010)

Introduction

Once scientific data has been gathered and interpreted, these results must then be communicated to people outside of the scientific community before they can be applied in any sort of public policy context. Of note, this communication of scientific results always takes place within a social context, and, inevitably, between scientists and non-scientist stakeholders as well as other members of the general public. Indeed even when scientific results are not applied to policy, they must still eventually be communicated in one form or another, very often to people who do not have scientific training themselves. Thus this activity of communication, beyond the research or laboratory context, is a vitally important part of being a scientist. Given this, it is somewhat surprising that scientists are rarely, if ever, trained in this process of communication with non-scientists during their graduate studies (something that should certainly be changed![1]). And yet, they are often still blamed – sometimes rightfully, and sometimes not – when this communication does not go well. However, it should be understood that just because the communication of scientific results between those who are trained as scientists and those who are not is sometimes difficult, as we will see in the example that follows, this does not mean that it is impossible to do well. While there are, of course, times when the communication of scientific results in the public setting goes awry, or is outright exploited for political gain, this has historically not been the norm, and so we should not be unduly concerned about or overly skeptical of the process, generally speaking. Nevertheless, the technical and moral difficulties that arise with science communication are indeed significant and thus should be acknowledged and addressed. For instance, as just one example,

DOI: 10.4324/9781003311072-4

these difficulties are very apparent regarding the issue of anthropogenic climate change. Despite overwhelming consensus amongst climate scientists regarding such change since at least the mid-1990s (Oreskes and Conway 2010), there is far from this level of agreement in the public realm. In June of 2021, for example, the *Economist* reported that one-third of Americans denied that human-caused climate change exists. This example (which we will return to later in the chapter in more detail) raises the more general question of why non-expert members of the public often do not trust experts when it comes to the communication of scientific matters. While there are many reasons for why this kind of mistrust occurs, one reason is that scientists have, in some instances, brought this upon themselves by engaging in what is sometimes called "wishful speaking." According to John (2019):

> A scientist engages in wishful speaking when she makes a claim which is not well- established, and where her motivation for making that claim is the non-epistemic benefits that follow from others believing (or believing that she believes) that claim, regardless of its truth.

An example of this would be an epidemiologist asserting to the public that a vaccine has no side effects when there is not sufficient evidence to justify this claim, simply because she knows that individuals are less likely to get vaccinated if they know there actually are side effects that come from the intervention in question (John 2019). In other words, wishful speaking sacrifices the truthfulness of a statement in order to promote or coerce a certain outcome in those who hear it. While John (2018) and others have suggested that avoiding openness and transparency in scientific communication in this way can actually be a good thing, because it can be used to motivate the public to act in desirable ways, there are at least a few obvious and significant problems with this sort of hardline consequentialist approach to science communication. First, such an approach is potentially morally problematic, since it constitutes a form of lying (Schroeder 2019). While a detailed discussion of the question of why lying is bad, in general, is beyond the scope of this chapter, one thing that we can say in defense of truthfulness (and thereby against lying) is that it facilitates autonomy – if people do not know the truth, then this inevitably places limits on their ability to act autonomously. Autonomy, simply defined, is "self-governance," or the ability to make decisions for oneself. Another way of stating this is that the term autonomy "typically designates an ability to […] to direct one's own choices and behaviors based on deliberation and reflection" (Racine et al. 2021). And this process of deliberation is hampered when incorrect information is what is being analyzed. Indeed, this is the justification generally given for the policy of informed consent in medicine – if a patient is not adequately (and accurately) informed about the proposed treatment or intervention, then he or she cannot truly consent to it. So if we value freedom of action, as well as self-rule, we should also value truthfulness.

Further, it is not entirely clear that lying to the public is even effective in producing the desired actions and outcomes that are intended when this is done. Instead, the opposite seems to be the case: people are often able to "see-through" this sort of dishonest communication, and, when they do, they tend to lose trust in the agencies or institutions that use it, and become even less likely to act on the recommendations that these institutions issue (Mathews et al. 2021; Malanowski et al. 2022).

In addition to this kind of attempt to manipulate *public* opinion, there are also some policy studies that "suggest that actors are influential when they 'frame' their evidence in simple, manipulative and/or emotional terms to generate *policymaker* attention" (Cairney and Oliver 2017, emphasis added). The idea is that these 'framing' strategies

> are effective because they appeal to the emotions and the familiar, combining facts with emotional appeals to prompt lurches of attention; telling simple and easily understood stories which manipulate people's biases, apportioning praise and blame and highlighting the moral and political value of solutions; and recognising the importance of interpreting new scientific evidence through the lens of the beliefs and knowledge of influential actors.

The idea, according to some, is that such paternalistic methods are justifiable, at least in certain cases, in order to achieve desirable ends, the worry being that:

> without this focus on the ways in which policymakers understand and respond to problems, scientists will be unable to exert influence, responding only to sudden policymaker demand for evidence-based solutions to a pre-defined problem.
>
> (Cairney and Oliver 2017)

However, others have voiced their concern that this sort of strategy is neither ethical nor practical, and has the "potential to reduce long-term scientific credibility" (Brownson et al. 2009; Cairney and Oliver 2017), and leave too much political power in the hands of science advisors. These are important concerns, and we will address them in more detail in the next chapter.

Another significant reason for public mistrust in science, at least in the United States, is due to misinformation campaigns that have been funded by industries that stand to suffer financial loss that would occur if the public were to understand the actual state of the science regarding the issue in question. Once such infamous campaign, which was conducted in the 1950s, was successful in casting doubt on the link between smoking and lung cancer for many years (Brandt 2012). This example, and others like it, clearly shows that the way that science is communicated in the public realm *matters* because the scientific beliefs that people hold often influence the way that they act, both publicly and privately. In fact, very often, public debates over science aren't actually over the science, but over what policies will follow from the science. If a person doesn't believe that a face mask, for example, is an effective tool in preventing the spread of infectious disease, then they will likely balk at a rule that requires them to wear one. What this means is that, at the least, we should aim for the honest and transparent communication of scientific results – because this seems to be the best way to build trust between the scientific community, policymakers, and the public, and this trust is of utmost importance when to comes to the application of data to policy (Malanowski et al. 2022). Further, as we saw in our discussion of inductive risk in Chapter 2, in addition to transparency, the timing as well as the way in which data is communicated must be carefully considered when the information in question has practical implications, and, in particular, when these implications are "high risk." And in order to weigh whether or not a given situation is high risk, scientists must appeal to extra-scientific concepts, ideas, and values.

So, if the communication of science matters in the realm of public discourse and public policy creation, because it potentially influences the way that people, including policymakers, act (or don't act), how then can we be sure that we can we trust scientists, in general, to communicate their data to us in a reliable way and thus that this data can subsequently be helpful in reliably informing our policies and actions? After all, if we know that "scientists are not guided by logic and objectivity alone," (Singer 1991) and thus are subject to all of the influences and biases that all humans are subject to, how are we to prevent inappropriate rhetoric or even "propaganda" from arising within the context of communicating scientific results to the public? This is certainly a question that is worth asking. However, to begin, we should note that while it is true that the process of science is an inherently human endeavor, with all of the good and the bad that this entails, this should not cause us to become overly skeptical of all scientists and their intentions. For one thing, science has inbuilt safeguards that are meant to protect against rogue opinions and personal agendas. These safeguards include both peer review as well as the public availability of scientific data and results. These things can help to mitigate personal biases and/or agendas on the part of individual scientists. And yet, we might be still tempted to ask – what if a majority of scientists are wrong? That is, what if most of them are confused, or disingenuous, or otherwise mistaken in their beliefs? These questions are worth asking because, in the end, what we need to know is that we can trust scientists, on the balance, to communicate their results to us non-scientists both accurately and honestly. And yet, we should recognize that it is very unlikely that all, or even most, scientists are disingenuous or confused or mistaken in their beliefs. To see why, consider the following analogy. Imagine we were to ask the following questions: what if all the plumbers were wrong? What if the way that all or most plumbers install plumbing into buildings, for instance, isn't really the best way? What if they are doing it in some particular way just to make more money, or because they are too lazy to invest time into researching what would, in fact, be the best way? While this scenario is certainly a possible one, it seems very unlikely to happen, at least over the long term, because while we could imagine that *some* plumbers could be incompetent or corrupt, it is difficult to imagine that the *majority* of them would be these things – simply because, eventually, word would get out. Word would get out because the plumbing would eventually fail – it wouldn't do what it is supposed to do, or what it is expected to do. And the same goes for scientists and their data. Certainly some scientists, however few, are incompetent and others are no doubt corrupt. But this should not cause us to be concerned that all, or even most of them, are – otherwise the history of science would not be the one of success that it very clearly is. From the discovery of penicillin in 1928, to the moon landing in 1969, to the recovery of the first asteroid sample in 2021, these are just a few examples of the success of science from the last century alone. This kind of broad-sweeping success would not take place if most scientists were incompetent or corrupt. And yet, while the method of science does indeed have a long history of shining successes, both predictive and innovative, we should not assume that it is a method that provides us with absolute certainty – no human endeavor ever does. Instead, as we have already seen, what science does provide us with is well-supported beliefs:

> Scientific ideas must be supported by evidence, and subject to acceptance or rejection. The evidence could be experimental or observational; it could be a logical argument or a theoretical proof. (Oreskes and Conway 2010)

Science, then, is not about facts (whatever one conceives of those entities to be), but is instead about *evidence* and its proper evaluation. In fact, the much-purported fact/opinion divide is, I believe, not only unhelpful to discussions of scientific communication and public policy specifically, but is actually also just plain false. Any claim – whether scientific or otherwise, in itself, should not be thought of as falling in to the category of either a "fact" or an "opinion" but rather as either into the category of a well-supported, or a not so well-supported claim. In other words, in any given domain, there are claims, and then then there are reasons, or evidence, that support, or "back up" these various claims. As we already saw in detail in Chapter 2, these reasons can be strong or weak, and they come in different kinds (such as empirical or normative). These reasons can and should be examined when we need to evaluate the evidence for the claims in question, but this is a different endeavor than attempting to categorize claims or statements into "facts" vs. "opinions" – which is a generally unhelpful, and at many times also misleading, exercise. Instead, our sole focus in the evaluation of claims should be on the evidence that supports them, in whatever form that takes, scientific or otherwise: claims that are well-supported should be accepted, and those that are not should be rejected, at least until there is further evidential support to back them up.

In addition to this, well-supported scientific claims, prior to becoming an accepted part of the body of general scientific knowledge must first

> be judged by a jury of one's scientific peers. Until a claim passes that judgment – peer review-it is only that, just a claim. What counts as knowledge, on the other hand, are the ideas that are accepted by the fellowship of experts.
>
> (Oreskes and Conway 2010)

Peer-review is thus an essential component in the generation of accepted scientific knowledge. However, at the same time we should be careful not to over privilege this process or to put too much weight on expert opinion, such as peer-review, over and above information gained from experiment, theory, or observation. Instead, we should understand the peer-review process as something that is a valuable "check" upon the activity of deriving scientific knowledge:

> Research produces evidence [...]. After that point there are no "sides." There is simply accepted scientific knowledge. There may still be questions that remain unanswered- to which scientists then turn their attention-but for the question that has been answered, there is simply the consensus of expert opinion on that particular matter. That is what scientific knowledge is.
>
> (Oreskes and Conway 2010)

Thus, we can conceive of scientific knowledge as being comprised of claims that are supported by logic, experiment, theory, and/or observation *and* that have passed the test of peer review. One or the other of these requirements, on its own, is not enough. But why not? In principle, at least, one would think that the methods of logic, experiment, theory, and/or observation *should* be enough to produce scientific knowledge on their own, because, after all, these methods, and the results that they produce are not (and should not) be determined by popular vote or expert opinion. Because of this, one might question the value of adding peer review to the scientific process at all. However, because the practice of science is an inherently human endeavor (Bendiscioli 2019), and subject to all that human endeavors of various sorts entail, the addition of peer review is helpful as a "check" to ensure that the process has been conducted properly – at least as far as is possible. In other words, via the process of peer review, scientists are able to control for, or at least mitigate, error, fraud, or bias on the part of their peers. However, peer review is not perfect, of course (and reviewers can also be disingenuous or biased). As Smith (2006) puts it: "Famously, it [peer review] is compared with democracy: a system full of problems but the least worst we have." While this may be true, we can still consider peer review to be a step that helps us to control for (and potentially eliminate), rogue opinions or nonscientific ideas that might otherwise pass as science in the public realm.

But all of the preceding discussion still leaves open the question of how we should make sense of genuine scientific disagreement, when it does occur. While we should be careful not to exaggerate either the magnitude or the frequency such disagreement, we should at the same time acknowledge both that it does in fact occur, and that genuine scientific agreement is a vitally important part of the scientific process, as it often drives discovery and/or spurs innovation. Thus, it is often a good thing! However, to see this, we must first understand scientific disagreement within its proper context. To begin, scientific disagreement is not the same thing as, for instance, political disagreement:

> While the idea of equal time for opposing opinions makes sense in a two-party political system, it does not work for science, because science is not about opinion. It is about evidence. It is about claims that can be, and have been, tested through scientific research-experiments, experience, and observation -research that is then subject to *critical review by a jury of scientific peers*. Claims that have not gone through that process- or have gone through and failed- are not scientific, and do not deserve equal time in a scientific debate.
>
> (Oreskes and Conway 2010)

In other words, scientific disagreement is not about personal opinion or preference, but instead occurs, for example, when experimental results conflict or when two or more theories or models are in tension with one another. When this happens, more research is generally called for in order to settle the disagreement. But again, in such a case, the disagreement is not personal, nor does it generally occur between individual scientists. However, personal disagreement over scientific results is something we should be skeptical of. As Oreskes and Conway (2010) have well-documented, there are some who would try to exploit minor scientific disagreements or uncertainties, or worse, invent some that do not exist, for political advantage. And while:

> Everyone is entitled to an opinion [...] when a scientist consistently rejects the weight of evidence, and repeats arguments that have been thoroughly rebutted by his colleagues, we are entitled to ask, what is really going on?

The question then is how we can determine when scientific disagreement is genuine vs. merely personally or politically motivated or disingenuous for some other reason. To attempt to answer this question, it will be helpful to look at an example in more detail.

Example: human-caused climate change

As an example that illustrates this issue of genuine vs. manufactured scientific disagreement, and the political and policy problems that the latter can cause, let's look again at the issue of anthropogenic (human-caused) climate change. To begin, it is certainly no secret that this issue is a highly politicized one, and has been for a long time. Back in 2006, a public opinion poll in *Time* magazine found that only 56% of Americans thought that global temperatures had risen – despite the fact that virtually *all* climate scientists thought so, and had for two decades or more. Even now, although the poll numbers regarding climate change have shifted some, they have not changed much:

> This denial of climate science may be expressed in a variety of positions: that the climate is changing; that the changes are anthropogenic, or caused by human activity; and that the results will cause significant harm to human populations, other species, and the earth's ecosystems. Each of these claims illustrates a variety of ways in which climate science denial may be manifested, but they hold in common a rejection of the clear consensus among scientists about what is known about climate change.
>
> (Mason 2020)

And to further complicate matters regarding this issue, "people's views about climate scientists vary strongly depending on their political orientation" (Funk 2017). For instance, the Pew Research Center recently reported that:

> The share of Americans who say global climate change is a major threat to the well-being of the United States has grown from 44% in 2009 to 60% this year (2021). But the rise in concern has largely come from Democrats. Opinions among Republicans on this issue remain largely unchanged. About nine-in-ten Democrats now consider climate change a major threat to the nation, that's up 27 percentage points from a 2009 survey. By contrast, the 6 percentage point increase among Republicans and Republican-leaning independents since 2009 is not statistically significant. In the new survey, about three-in-ten Republicans (31%) consider climate change a major threat, while 45% say it is a minor threat and 24% say it is not a threat to the nation.[2]

Further, in June of 2022 the conservative-majority Supreme Court of the United States:

> limited the Environmental Protection Agency's ability to prevent power plants from releasing climate-warming pollution. The court ruled that Congress had not given the agency the authority to issue the broad regulations that many climate experts believe could make a major difference — the kind of regulations that many Biden administration officials would have liked to implement.[3]

The ruling is the latest sign that the Republican Party in the United States is unconcerned about climate change. The six justices in the majority were all Republican appointees; the three dissenters were all Democratic appointees.

Adam Liptak, The Times's Supreme Court correspondent, wrote

> Chief Justice John G. Roberts Jr., writing for the majority, only glancingly alluded to the harms caused by climate change. Justice Elena Kagan began her dissent with a long passage detailing the devastation the planet faces, including hurricanes, floods, famines, coastal erosion, mass migration and political crises.[4]

There are at least two interesting philosophical issues worth examining here. The first is the question of why there is disagreement between climate scientists and the public on this topic at all, and another is the question of why this disagreement/denial regarding the state of the science breaks down along sharp political lines, at least in the United States.

In reference to the first question, some have argued that the public's (and thereby policymakers') denial of human-caused climate change can be explained by science illiteracy, lack of education and/or simple reasoning errors (Coady and Corry 2017). However, on closer examination, this does *not* seem to be the case. Instead, the denial of human-caused climate change by a large segment of the American public rather seems to be due to a *mistrust* of climate scientists and the data they report. In fact, according to Kahan et al. (2012), greater analytical skill was not correlated with attitudes in support of anthropogenic climate change. Instead, advanced scientific and mathematical abilities were predictive of polarization, which is the intensity of one's commitment to one's position, rather than to the position itself. Interestingly, this phenomenon is not limited to the United States:

> while expert and lay opinion were more closely aligned in Europe than in the US, lay/nonexpert Europeans were similarly unlikely to be swayed by expert opinion. Second, the researchers find a pattern of polarization. After being told of consensus among experts, respondents with an interest in public policy or economics were more likely to change their opinions to match those of experts, but respondents without those interests, or who had previously indicated they distrusted academic expertise, moved in the opposite direction.[5]

In other words, it's not the case that lay members of the public lack the education needed to understand the climate science data that has been communicated to them, but rather that they do not believe it to be accurate as reported. That is, they simply do not trust what they have been told. But the question then is, "why not?" Why is there disbelief in this particular case? We did not, for instance, see this sort of widespread mistrust or disbelief when NASA landed a planetary rover on Mars, or when medical scientists transplanted the first pig heart into a human heart. What, then, is the difference in regard to climate change? Not surprisingly, the short answer to this question is money. To elaborate further, the significant difference in the climate science example is that in this case there was a calculated and targeted misinformation campaign in the United States that was aimed at purposely confusing public opinion on the matter. This campaign, which was organized by the fossil fuel industry, which stood to lose financially if the true state of the science were to become widely known and accepted, sought to cast doubt on the veracity of climate science data by convincing the public that these data were not agreed upon, but were rather unsettled and still under debate within the scientific community. Disturbingly, this campaign was highly effective:

> Two crucial developments during the presidential campaign of 1988 changed climate science forever. The first was the creation of the Inter-governmental Panel on Climate Change. The second was the announce-ment by climate modeler James E. Hansen, director of the Goddard Institute for Space Studies, that anthropogenic global warming had begun. An organized campaign of denial began the following year, and soon ensnared the entire climate science community.
>
> (Oreskes and Conway 2010)

> The Intergovernmental Panel on Climate Change published its first assessment of the state of climate science in May 1990. It reiterated the result that was by now familiar to anyone who had been following the issue: unrestricted fossil fuel would produce a rate increase of global mean temperature during the next century of about. 3 degrees C per decade; this is greater than that seen over the past 10,000 years. Global warming from greenhouse gases would produce changes unlike what humans had ever seen before.
>
> (Oreskes and Conway 2010)

However, in a concerted effort to stir up controversy, the conservative-leaning Marshall Institute, which was originally established in order to defend President Reagan's Strategic Defense Initiative against attack by other scientists, issued a statement which said that this effect was not caused by humans, but could rather be attributed to solar variability. Even though the IPCC subsequently addressed – and rejected – this claim from the Marshall Institute, it didn't seem to matter, and the seeds of doubt that climate change was not actually human-caused were sown. To be clear, the question at that time was not about whether or not climate change existed – all parties agreed on this – but about whether or not it was primarily caused by humans or by other factors. While a body of scientific evidence that it was (and is), in fact, caused by humans began to accumulate, the Marshall Institute continued to argue that it was not. These claims were:

> Taken seriously in the Bush White House and published in the *Wall Street Journal,* where they [were] read by millions of educated peo-ple. Members of Congress also took them seriously. Proposing a bill to reduce climate research funding by more than a third in 1995, Congress-man Dana Rohrbacher called it "trendy science that is propped up by liberal/left politics rather than good science."
>
> (Oreskes and Conway 2010)

This sort of politicization of science occurs in other countries as well. For example, in Australia, climate science debates are also centered around the perception that scientists do not agree with one another on the state of the science. This what is known as a "consensus gap."

> A consensus gap is the difference between the public's perception of how much agreement there is among scientists that humans are causing global warming (typically about 50%), compared to the actual 97% consensus among scientists publishing in the peer-reviewed literature.[6]

An examination of this historical background in both the United States and elsewhere reveals that financial motivations[7] turned climate science into a serious political fight, one that has persisted for three decades, both in the United States as well as other countries abroad.

This now leaves us with the second question – that of why this mistrust of climate scientists breaks down along sharp political lines. First of all, it is interesting to note that this political divide is not found in all areas in which there is public mistrust of science. When it comes to the issue of vaccine hesitancy, for example, (aside from COVID-19 vaccine hesitancy, which is a special case), this mistrust is equally distributed amongst liberals and conservatives in this country. In 2015, the Pew Research Center conducted a survey[8] of two thousand adults which concluded that approximately 12% of liberals and 10% of conservatives in the United States believed that childhood vaccines are unsafe. Since then, other studies at the University of Idaho, Yale, and Fordham have produced similar results. Further, the latest vaccine exemption data from the CDC reports that liberals and conservatives are equally likely to request such exemptions.

This is likely due to the fact that those who identify as politically conservative are often distrustful of government agencies, such as the CDC, and thus skeptical of vaccine recommendations that come from them, while those who identify as politically left-leaning are often distrustful of the pharmaceutical companies who manufacture and sell the vaccines and thus often hesitate to vaccinate themselves or their children for this reason. On both sides of the political spectrum, those who have a higher level of education (especially those who hold graduate degrees) are more likely to be vaccine hesitant than those who do not.

At the same time, there are factors other than political party and ideology that shape the public's often-complex views on science matters such as vaccination. For instance, there are notable issues on which racial and generational differences are pronounced, separate, and apart from politics.[9]

All of this shows us that:

> The scientific enterprise is complex and so, too, is public opinion about science. The notion of trust itself has multiple dimensions. Public

trust in scientists encompasses expectations about scientists' actions, trust in scientists to be honest brokers of information, trust in scientific expertise and understanding, and trust in the motivations and influences operating on science research. Viewed through that lens, levels of public trust in science are quite varied, particularly across scientific domains.

(Funk 2017)

And indeed, these political leanings matter, because in some cases, such as with human-caused climate change, the lack of public trust in scientific consensus can be particularly disheartening, especially because it can delay or prevent actions that could make a difference to the – in this case very significant – problem at hand. And thus while it is clear that, "on balance things would be better if the public placed greater trust in science and scientists, at least on certain issues" (Schroeder 2019), the lack of trust in this one particular area should not cause us to completely lose heart in the process of scientific communication altogether, as this mistrust does not extend to all areas of scientific research.

Further, it is also encouraging to note that

public trust in scientists is stronger, by comparison, than it is for several other groups in society. For example, many more people report trust in information from medical scientists, climate scientists, and GM food scientists than information from industry leaders, the news media, and elected officials.

(Funk 2017)

Further, a Pew Research Center survey found in 2019 that 86% of respondents trusted scientists at least "a fair amount." So while there is, in fact, a problem of public mistrust of science in some areas of research, it is not an all-pervasive (or intractable) problem.

One might then reasonably ask what can be done to increase public trust in science when it comes to the climate and to other controversial issues like it. As the start of a potential answer to this question, we can learn some important lessons in this area by examining in more detail the problem of vaccine hesitancy. Maya Goldenberg (2016, 2021) has written insightfully on this topic. She makes the convincing claim, that while "public resistance towards scientific claims regarding vaccine safety is widely thought to stem from public misunderstanding (or ignorance) of science," in actuality the problem is one of mistrust of scientific experts and institutions. In this way, we can see that this problem is similar to the climate science issue: the problem is not that the public doesn't understand the data, the problem is that they do not trust it.

What, then, can be done to rectify this lack of trust in scientists on the part of the public? First, if the problem is not one of ignorance, then education campaigns that treat it as such clearly won't work – and can even backfire if they are perceived as pedantic. And, in fact, at least in the higher income countries, a higher level of education is correlated with a greater likelihood of refusing vaccination (Bergen et al. 2023). This is likely the case because those with more education are less likely to trust what people in positions of power or authority say, at least without looking into the issue for themselves.

So if education is not the answer (because ignorance is not the problem) to solving the problems of climate denial or vaccine hesitancy nor is lying (because it is both immoral and ineffective) then what is? According to Goldenberg, instead of unhelpfully framing the conflict as one of science vs. ignorance (Fischer 1996), scientists should instead focus on open and honest communication with the public, as well as a willingness to conduct further research:

> Specifically, vaccine hesitators want investigation into the admittedly rare but serious adverse events that they associate with vaccines. The main stream insistence that, to quote the Health Canada (2011) brochure, "it is often very difficult to determine if a 'reaction' was directly linked to a vaccine or was an unrelated 'event' which would normally occur in a population," is grounds for further research rather than secondary to the overall social benefit that vaccination programs provide.
>
> (2016)

Regarding communication, then, Goldenberg's argument is directly opposed to John's: instead of lying to the public in an effort to motivate action, we should require openness and honesty on the part of scientists, in order to foster trust. Regarding the issue of further research, in 2015 I argued along similar lines:

> underneath the highly political, sensational and divisive surface of [the vaccine] debate is a deeper ethical and evidential issue: we simply do not have enough research data to resolve this issue. And until we address this research gap, the debate will continue.
>
> Of course there is a lot of research that we *do* have. We know that vaccines are effective against the acute communicable diseases that they are designed to prevent. We also know that, *for the most part,* these vaccines pose little medical risk to those who receive them (although whether or not they are harmful to certain sub-populations is still not well understood). But what we don't know is how and whether vaccines affect the risk of certain chronic *non-communicable* diseases

(NCDs) and conditions such as asthma, allergies, autism spectrum disorder, multiple chemical sensitivity, and irritable bowel disease, all of which have been on the rise in highly vaccinated countries in recent decades.

We don't know anything about the relationship between vaccination and NCDs because we haven't done any studies that compare vaccinated groups against unvaccinated groups. While such studies are the standard for approval for all other medical interventions, comparative studies that check for NCD harm from vaccines are non-existent. This is due to the pro vaccine stance that such trials in the case of vaccines would be unethical (because it's not ethical to leave anyone unvaccinated).

In particular, while it is known that vaccines affect the microbiome (just as antibiotics do), which in turn affects immune responses, and causes non-specific side effects, we don't know *how* vaccines have changed our internal flora, what these side effects are, and whether or not these changes in turn have affected the prevalence of NCDs and conditions. In other words, it is not known whether or not vaccines, while preventing acute, communicable diseases, in turn increase the risk of certain chronic, non-communicable conditions and diseases. In order to learn how and whether vaccines affect chronic health conditions, we need to begin systematically collecting NCD data in connection with vaccine schedules in both developed and developing countries.

If we want to resolve the debate between the pro- and anti-vaccinationists, then we must move beyond the political and commit to conducting comparative studies. In order to "let the evidence decide," we must first collect the data.

(Kennedy 2015b)

To summarize, what this means is that when further research is truly warranted, we need to be willing to pursue it, for the sake of fostering public trust in science. This is not to say that further research is required every time a disagreement over data is voiced by one or more members of the public. Instead, the requirement that must be met is that there should be a state of genuine uncertainty *within the scientific community* on the matter in question. If there is, then further research is warranted.

Improving science communication

This two-fold approach to science communication: honesty and transparency when reporting data, coupled with a willingness to conduct further research when it is genuinely required in order to resolve uncertainty, can go a long way toward improving public trust in scientists and scientific institutions. However, to effectively do these things, it is essential that scientists first begin by treating non-scientist stakeholders, as well as other non-scientist members of the general public, as partners rather than as adversaries. In particular, an "us" vs. "them" approach is ultimately unhelpful in building public trust in science, especially when it assumes that non-scientists are always ignorant, misinformed, or generally less intelligent than their science-trained peers, which, given the research we have available, is clearly not the case. This is one of the significant problems with the sort of paternalistic approach advocated by John (2018): it assumes that the public is too ignorant and/or unsophisticated to understand explanations about scientific uncertainty:

> Unfortunately, just as publicising the inner workings of sausage factories does not necessarily promote sausage sales, so, too, transparency about knowledge production does not necessarily promote the flow of true belief throughout the population (and so on for honesty, sincerity and openness).
> Being 'honest' about uncertainty may help clarify which claims are well established. However, these norms can backfire: openness and transparency may [...] lead hearers to reject 'well-established claims'.

In particular, one issue with John's account, and others like it, is that it assumes that there is a sharp divide between "the public" and "the experts" in the first place. But this isn't necessarily the case. For example, we might reasonably wonder into what category we would place philosophers or ethicists – would they count as members of the general public, from whom some information should be hidden? Or would they count as (epistemically privileged) experts? These questions are not explicitly answered in John's paternalistic account. It seems, however, that if the two assumptions of (a) public ignorance and (b) a sharp public vs. expert divide lead us to a policy of hiding the truth in order to force public action – then these assumptions are a mistake. Instead, we should encourage scientists to engage in what is known as "epistemic humility." Epistemic humility (Schwab 2012) is a characteristic of claims that accurately portray the quality of evidence for believing the claim to be an accurate one. In other words, epistemically humble claims neither overreach nor under-reach (nor over – nor understate) the evidence that supports them. Engaging in epistemic humility, in turn, facilitates better communication between scientists and non-scientists. This is because being epistemically humble encourages the acknowledgment of uncertainty, when it is present, which then sets the stage for open and honest communication.

Some might object that taking the necessary time to adequately explain scientific uncertainties to non-scientist would require undue effort. Others worry that it might be dangerous. For example, when it comes to vaccine science:

> [Many] scientists are so terrified of the public's vaccine hesitancy that they are censoring themselves, playing down undesirable findings and perhaps even avoiding undertaking studies that could show unwanted effects. Those who break these unwritten rules are criticized.[10]

However, as we have seen, this fearful hiding of data from the public only tends to backfire in the long run (because people eventually figure out that they have been lied to) and thereby foster further distrust of scientists and public institutions. Further, it seems that putting in the effort required to explain scientific uncertainty, when it is present, is likely to be well worthwhile, and far preferable to the alternative, because failing to acknowledge and/or communicate scientific uncertainty can have seriously detrimental consequences. Not only can hiding uncertainty, when it is present, eventually lead to the erosion of public trust, it can also affect the way that science itself is practiced. For example, ignoring uncertainty in medical cases can lead to what is known as "premature closure," which is the cessation of a diagnostic investigation before it is appropriate. To see this more clearly, consider the following case study:

> A 41-year-old man who worked in construction presented to our psychiatric clinic complaining of depressed mood that had started two months previously. He related his condition to the recent and violent loss of his spouse, who was killed in a car accident. The patient's symptoms included loss of concentration, lack of sleep, and loss of appetite leading to a six-pound weight loss over the previous two months. He denied any suicidal thoughts, but admitted that he sometimes heard his spouse's voice in the house. After discussing the condition as a case of depression with the patient, we started the patient on psychotherapy sessions (twice monthly) and 20 mg of fluoxetine daily. The patient came to the clinic for follow-up two weeks later. He complained that the therapy had done nothing to improve his mood, and that he thought, in fact, his condition was worsening. He stated that two days previously, he thought he saw his deceased spouse in the kitchen. He also reported having thoughts about death and the futility of life without his partner, although he insisted he was not suicidal and that he had no plans for committing suicide. We advised the patient to continue taking the fluoxetine and we increased his psychotherapy sessions to weekly. After this follow-up visit, the patient, however, only attended one psychotherapy session and was lost to follow-up for several months. Four months later, the patient presented to our emergency room (ER) with delirium, mild fever,

and visual and auditory hallucinations. According to the ER team, the provisional diagnosis was substance abuse, but the toxicology screen came back negative, denying this diagnosis. The treatment team noted a weak, thready pulse, severe hypotension (70 systolic), and severe hyponatremia and hyperkalemia. A diagnosis of an Addisonian crisis was made. The patient was admitted to the intensive care unit (ICU), and proper intervention for the Addisonian crisis was administered. Later, it came to our attention that the patient had visited several physicians and hospitals in the area over the past four months, but a diagnosis of Addison's disease was not made by anyone.

(Abdel-Motleb 2012)

In this case study, the patient in question nearly died because the uncertainty in his initial diagnosis was not acknowledged and therefore was not properly investigated or addressed. The same can happen when uncertainties are mishandled in climate science, vaccine research, or any other area of scientific inquiry: when uncertainty is present, it must be both acknowledged and truthfully communicated in order to practice science well. Ignoring uncertainty, and then, because of that, stopping investigation and/or experimentation too soon, can lead to detrimental consequences including, but not limited to, poor application outcomes as well as the erosion of public trust in the scientific community. Thus, scientific uncertainties should not be something that we are afraid of, or find ourselves tempted to ignore, but should instead should be viewed as opportunities to fuel further research, investigation, and inquiry – precisely because the only way to effectively eliminate uncertainty is via an appeal to the appropriate amount of supporting evidence. In practice, this means that all scientists should be taught to acknowledge and also to communicate uncertainties in their research *especially* in controversial areas, such as climate science, because, in the end, this is the most likely way to (a) ultimately resolve the uncertainties when necessary and (b) build public trust in scientists as well as the method of science more generally.

Conclusion

To conclude, building public trust in science is important for a variety of reasons, one of which is that without an adequate level of such trust, people will be unlikely to listen to or to follow science-based policy recommendations. Further, as we have seen in this chapter, the way to build this trust is not by attempting to educate the public when ignorance is not the problem, nor by lying to them in an effort to coerce their actions, but instead by engaging in open and honest communication and dialogue as well as conducting further research when it is warranted. It should be made clear, however, that this does not mean that we should think that scientists and non-scientists are on equal epistemic footing when it comes to scientific matters: clearly, they are not (otherwise what would be the point of scientific training?). Scientists are, by definition, experts in their fields of research and as such have more knowledge within their various domains of scientific inquiry than those who don't have this training. This means that, as a general rule, if the scientific community and the general population of non-scientists disagree on the particularities of a scientific matter, then we ought to privilege the scientific community's view. This, of course, does not mean that scientists cannot ever be wrong. Clearly they can be. However:

> if you ask whether a group of well-trained researchers, thoroughly familiar with the details of the issues [...] will be *more likely* to be right than an uninformed public, the answer seems obvious: even if you cannot be sure, you know where to place your bets.
>
> (Kitcher 2011)

Yet, even given their unequal epistemic positions, scientists and non-scientist members of the public can and should engage in open, honest, and respectful dialogue with one another concerning scientific matters. We will turn to a further discussion of this topic, and why it is important in the context of scientific expertise, in the following chapter.

Notes

1 Master's degrees in science communication and science journalism are becoming increasingly popular, undoubtedly at least in part because the need for the effective communication of science in the public sphere is being recognized. However, most PhD programs in science do not include this sort of training, and so most scientists do not receive it.

2 https://www.pewresearch.org/fact-tank/2020/04/16/u-s-concern-about-climate-change-is-rising-but-mainly-among-democrats/.

3 https://www.nytimes.com/2022/07/01/briefing/supreme-court-epa-ruling-climate.html

4 https://www.nytimes.com/live/2022/06/30/us/supreme-court-epa?campaign_id=9&emc=edit_nn_20220701&instance_id=65512&nl=the-morning&regi_id=117975981&segment_id=97324&te=1&user_id=ab836b623b8275f19ee5354a015ccb4f.

5 https://www.chicagobooth.edu/review/downfall-and-possible-salvation-expertise.

6 https://psychology.org.au/community/advocacy-social-issues/environment-climate-change-psychology/resources-for-psychologists-and-others-advocating/the-psychology-of-climate-change-denial.

7 Money is not the only thing that can cause an area of scientific inquiry to become controversial, however. Any research that challenges deeply held personal beliefs and/or values can also do this. For example, consider the research that showed that racial categories are not genetically determined, as was once

thought to be the case (Rosenberg 2002) and thereby challenged the longstanding "belief that racial and ethnic groups have substantial, well-demarcated biological differences and that these differences are important" (National Human Genome Research Institute 2005). There were some who did not (and still do not) want to accept these research findings because they do not cohere with their pre-existing and deeply entrenched worldview.

8 https://www.pewresearch.org/science/2015/07/01/americans-politics-and-science-issues/.
9 https://www.pewresearch.org/science/2015/07/01/americans-politics-and-science-issues/.
10 https://www.nytimes.com/2018/08/04/opinion/sunday/anti-vaccine-activists-have-taken-vaccine-science-hostage.html.

5   Scientific expertise

> Scientific conclusions – theories, concepts, facts – are enormously useful for individual and political decision making, but only if they are regarded as that: tools for thinking (and not as commands for action).
>
> (Reiss 2019)

Introduction

As we saw in the previous chapter, before we can begin to think about how to apply science to public policy, we first need to understand how to communicate it well to the public and to other stakeholders. In this chapter, we will discuss the related topic of scientific expertise, which is closely connected to scientific communication, in the following way. Because it is often scientific experts (more on who counts as an "expert" will follow later in the chapter) who are tasked with the responsibility of communicating scientific results to the public and to other stakeholders, we need to know who the experts are. To put it another way, we should care about who the scientific experts are, because we need to be able to identify them if we are going to rely upon them to give us accurate and applicable information. Yet, scientific expertise, and indeed the definition of who counts as an expert at all, in any field, has become increasingly controversial in recent years. There are several reasons for this, but perhaps the most significant one is that the question of expertise and the issue of trust are closely connected: no one wants to listen to, or to follow recommendations given by, any "expert" that they do not trust. And this is understandable: we have to be able to trust our experts in order to know that we are getting accurate information from them that we can then aptly apply to our public policies and/or our individual actions. On the one hand, some people tend to "under-trust" experts, while, on the other hand, others tend to "over-trust" them. As we will see in what follows in this chapter, neither situation is ultimately helpful when it comes to policy decisions. Perhaps all of this seems

DOI: 10.4324/9781003311072-5

obvious, to you, but it is not obvious to everyone. Consider, for example, that it is becoming increasingly common to hear people say that members of the public[1] should simply blindly trust the "experts," particularly scientific experts, when it comes to public policy decision-making:

we are told to submit to the authority of science (including social science) and to leave technical questions, including technical questions of great social relevance and potential impact, to the scientific experts — because they know what they are talking about.

(Reiss 2019)

And this is done without any appeal to who counts as an expert, or instructions on how to go about identifying one. The idea seems to be that an expert, in any given domain, is qualified not only to convey accurate information to the general public, but also to prescribe, or recommend, actions to that public. In particular, this kind of appeal to scientific expertise has recently been made in many countries around the world during the ongoing COVID-19 pandemic. For example, in the United Kingdom,

the government's response to the pandemic has been consistently presented to the public as "following the science." Daily press conferences began on March 12, 2020 with a briefing delivered by the prime minister flanked either side by the government's chief medical officer, Chris Whitty, and chief science officer, Patrick Vallance. Vallance and Whitty have since played a prominent role in communicating the government's policy on the coronavirus, standing alongside government ministers in many more press conferences and appearing on television, on radio, and in print to explain the UK's public health strategy.

[...]

This "following the science" message depends on the public believing not just that these scientists are part of the communication of the policy, but also that they are involved in deliberations that result in that policy. That the UK government is in fact following the science is very much open to question, but it is evident that the UK's approach to communicating its message to its citizens is to heavily emphasize its (putative) technocratic nature.

(Bennett 2020)

Similarly, in the United States, Dr. Anthony Fauci, who was the director of the U.S. National Institute of Allergy and Infectious Diseases and the chief medical advisor to the president, regularly appeared alongside the American president during public coronavirus briefings. The sometimes implicit, sometimes explicit, message was that he, as an expert, represented "the science" and thus was qualified to tell the public not only what was the case regarding the pandemic (such as, for example, how the virus is transmitted, or the number of cases in any given state or region, or the estimated effectiveness of COVID-19 vaccines against various strains of the virus) but also what to do about, or with, these facts. Further, there is also a more worrying implicit claim that is often made, which is that scientific "experts," such as Fauci, are somehow beyond reproach, simply in virtue of being scientists. For example, in June of 2021, Fauci remarked in a television interview[2] that "a lot of what you're seeing as attacks on me quite frankly are attacks on science, because all of the things that I have spoken about consistently from the very beginning, have been fundamentally based on science," essentially arguing that because he is a scientist, he could be wrong and therefore should not be "attacked."

There are all sorts of problems with this claim, not the least of which is that, as we have seen, science is a method that, albeit reliable, is certainly not infallible, nor does it have an unlimited domain. In short, this kind of increasingly common and widespread appeal to and promotion of the purported epistemic and normative authority of the scientific expert raises several interesting philosophical questions, which we will examine closely in this chapter. First, it begs the question of who counts as a scientific expert. Next, it also raises the issue of whether there is a connection between scientific expertise and epistemic authority, as well as the question of what, if any, connection there is between scientific expertise and normative, or moral, authority, particularly in the domain of public policy. Thus in this chapter I will make the case that while we do have reason to trust scientific experts to give us accurate scientific information, this (alone) does not qualify these experts to prescribe actions to the general public. Instead, it takes more than scientific expertise to undergird the moral authority to prescribe action outside of the scientific domain.

Scientific expertise and epistemic authority

Before addressing the question of whether or not scientific experts have either epistemic or normative authority (in virtue of being experts) we first need to know, at least roughly, what an expert[3] is – and this question is far from being a settled one. However, here I will adopt a simplified view, building on Goldman (2001), Croce (2019), and Bennett (2020) of what it means to be an expert in a given domain, and then extending this view in order to propose a definition of what it means to be an expert in the public domain, in particular.

According to Goldman, we can understand an "expert" to be someone who possesses more accurate information (that is, someone who has more true beliefs⁴) than most people do in a given domain. On this definition, then, there is a situation of epistemic asymmetry between someone who is an expert and someone who is a non-expert in any given domain, a distinction which is sometimes known as the novice-expert dichotomy. Thus, a scientific expert on this definition is someone who knows more (or at least has more true beliefs) about some scientific subfield than most people do. However, some have argued against this view that merely having more true beliefs in a domain is not enough to constitute expertise, suggesting, instead, that we need to add (Croce 2019) to this the requirement that an expert must also understand, and be able to explain, his or her beliefs to others,⁵ while citing evidence in their given field that supports these beliefs (Walton 1989). In other words, this added requirement is that the expert must have reasons for their beliefs *and* be capable of explicating these reasons to non-experts. This seems to be a reasonable requirement to add when we are talking about *public* experts specifically – i.e. when we are talking about experts who are displaying or employing their expertise in the public domain, to an audience of non-experts. On this enhanced definition, then, an expert is someone who is *competent* in their field, in that they both possess more true beliefs in the area of expertise *and* are capable of relaying accurate information about their beliefs in that field to others.

However, as some have pointed out, even competency, as described above, is alone not enough for expertise: just because someone is competent, that does not mean that they are reliable, and it is certainly the case that we want this, too, to be true of our experts – or at least of our public experts. Another way of saying this is that we want our public experts to not only be competent but also to be sincere (Bennett 2020). And we want them to be sincere, not simply because sincerity is a nice way for people to be, but because insincerity and unreliability often go hand-in-hand. If someone isn't sincere, the information that they relay is not likely to be reliable. Given this concern, it seems reasonable to define a public expert as someone who is "epistemically trustworthy,⁶" in that they are both competent (that is, they have more true beliefs than a non-expert, and are capable of explaining these beliefs to others) and sincere. This, in turn, means, to put it simply, that an insincere "expert" really isn't an expert at all.⁷

Having now defined (at least roughly) what it means to be an expert,⁸ we can turn to the question of why we ought to care in the first place about who counts as an expert. This is generally agreed to be because we think that there is a relationship between expertise and what is known as *epistemic authority,* where an epistemic authority can be understood to be someone who "can help their interlocutors achieve some epistemic goal in a given domain through their superior knowledge and/or understanding" (Croce 2019). Thus, the idea is that we should care about who the experts are if we have the goal of wanting to improve our own epistemic positions regarding some domain or some given set of particular questions within a domain, and experts are able to give us the information needed to do this – information that we cannot get on our own (in virtue of being non-experts). In other words, we care about who the experts are because experts are people that we can learn from.

If an expert then is someone who is both competent and sincere, and we are interested in identifying who these people are in order to gain more knowledge in a given domain, then we (obviously) need to know how to identify people who are competent in their fields and also sincere. This first requirement of competency is important because, as Winsberg et al. (2020) point out, even highly regarded "experts" are capable of making epistemic mistakes. They argue that:

> we have strong grounds in general to be skeptical about experts' predictions on hard problems. For instance, in *Expert Political Judgment*, Philip Tetlock (2005) examined nearly 83,000 predictions made by experts in a variety of fields. He focuses on what the experts themselves consider hard problems rather than easy problems. In general, he finds that on such questions, experts performed poorly, barely better than Berkeley undergraduates. Tetlock's work warns us against simply "deferring to the science" on hard predictions, since the science in fact shows the scientists are bad at such predictions.

While we might initially assume that this particular criticism could be deemed unfair, given that it rests on an assessment of the performance of experts only on expert-identified "hard problems," and that it could simply be the case that some problems are so difficult that no amount of training or expertise will make a difference when confronted with them, the general point holds: competency in an expert is something that it is reasonable to expect and thus it is important for non-experts to be able to have something to appeal to when making judgments regarding the competency of a purported expert. While this is not always easy to do, generally speaking, most people agree that indirect criteria such as degrees, track records, consensus statements, etc., are reasonable (but not infallible) as proxy for assessing this criterion.[9] For the most part, these indirect criteria are determined by the peers of the potential expert. That is, for instance, in the case of scientific expertise, we necessarily rely on other scientists to assess the standing of their peers as experts. And although it is imperfect (as are all forms of peer review), this system is generally agreed to be better than alternatives[10] (Gallo et al. 2016).

After assessing (as well as possible) the competence of a potential expert, it is then of equal importance to assess their sincerity, given that it is well known that some "experts" can be disingenuous, or worse. For example, early in the COVID-19 pandemic

> many health experts, including the surgeon general of the United States, told the public simultaneously that masks weren't necessary for protecting the health of the general public and that health care workers needed the dwindling supply of masks in order to stay safe.
>
> (Tufecki 2020)

Further reinforcing this message in a 60 Minutes interview on March 8, 2020,[11] Dr. Fauci stated that:

> There's no reason to be walking around with a mask. When you're in the middle of an outbreak, wearing a mask might make people feel a little bit better and it might even block a droplet, but it's not providing the perfect protection that people think that it is.

Then, just a few months later, and in the absence of any new data, the same health officials announced that masks were essential for everyone to wear in public settings in order to decrease the transmission rate of the SARS-CoV-2 virus. Again, this change in policy did not reflect any change in the science – there was no new data or new experimental information of any kind that became available in the interim between the time that the two messages were conveyed. Instead, the two messages were simply contradictory – either masks work (to some degree of efficacy) to protect people from the virus or they do not. And yet, this contradictory messaging was clearly and regularly conveyed to the American public during the early days of the pandemic. What happened subsequently, when the non-expert public saw straight through the (ridiculously) contradictory messaging, was that there was a public outcry from a subsection of the population who used it as proof that the "experts" – across the board – were not to be trusted. Or, perhaps worse, that there really is no such thing as an expert at all. But these kinds of views are, ultimately, untenable – no one can be competent in every domain, and thus it is imperative that we both be able to identify, and rely upon, genuine experts to inform us about topics and issues that we do not ourselves have expertise in. This does not mean, of course, that we ought to put "blind" trust in anyone, experts included, instead, what it means is that we need to be able to identify experts who are both competent and sincere, and thereby likely to be reliable.[12] And this is the reason why only competent, trustworthy individuals should be counted as experts.[13]

Scientific expertise and epistemic uncertainty

We have now defined a genuine expert as someone who is both knowledgeable in their field and credible, but this, of course, does not mean that they are thereby infallible or are able to report data with a 100% level of certainty. This is due to the fact that, in addition to all humans always being fallible, all scientific inquiry and all scientific data is also uncertain as well. In other words, as we have already seen, this means that some level of uncertainty is always present in every area of scientific inquiry from epidemiology to climate science to physics. And while no one really likes this fact – and we all wish that we could do away with scientific uncertainty entirely, this does not mean that science is a flawed method or that we cannot, eventually, aptly apply its results to our policies. But, of course, scientific uncertainty is neither easy to deal with, nor likely to ever be completely removed, even with continued advancements in knowledge and instrumentation. It seems that the best thing that we can do, then, is learn how to deal with, and how to communicate, this uncertainty. The first step in this process, after recognizing that uncertainty exists, is to ensure that the uncertainty in question isn't hidden by scientists and other researchers, but instead is acknowledged, and communicated, both to other scientists, as well as to the public, to policymakers and to other stakeholders more generally. Communicating scientific results to the public, especially when there is a high level of uncertainty, however, is often easier said than done. This is because:

> communicating scientific uncertainty *requires both simplifying and complicating* normal scientific discourse. On the one hand, the uncertainties that it addresses must be reduced to their decision-relevant elements. On the other hand, the uncertainties that scientists fail to mention must be uncovered. Which uncertainties to subtract and add depends on the decisions that the communications are meant to serve.
>
> (Fishhoff and Davis 2014, emphasis added)

Yet, open acknowledgment and communication of scientific uncertainty is the best way to handle it, because when uncertainty is *not* acknowledged and/or is improperly communicated, this can backfire: hiding the truth from "the public" serves to eventually only foster distrust of the "the experts." So, as Tufecki (2020) argues, it's better to simply tell people the "full painful truth," because trust is more likely to be fostered (and policies to be followed) when people recognize that they are being treated with respect. Of course, scientists are often aware of uncertainty in their research results, but are yet not able to quantify this uncertainty precisely – that is, the probability estimates of the level of uncertainty in any given data set are themselves often uncertain (Stanford 2007). This too can create problems when communicating scientific results to the public – particularly when the public might demand to know how "certain we are about being uncertain." To further complicate matters,

> The word "uncertainty" itself has slightly different meanings when used in everyday speech versus a scientific context. In scientific discourse, it conveys the degree to which something is known. In the vernacular, the word conveys rather the sense of not knowing. The difference is subtle, but important.
>
> (Scientific uncertainty 2019)

Scientific uncertainty, then, should not be understood as an epistemic state of complete lack of knowledge, but instead as a state in which the knower possesses knowledge to a certain degree. Indeed, this epistemic position applies not only to scientific knowledge specifically, but to all knowledge derived from inductive reasoning. Logically speaking, there are two main types of reasoning: deductive and inductive. In deductive reasoning, if the premises of the argument in question are true, then the conclusion of that argument is *guaranteed* to be true as well.

For example, the following argument:

Premise 1: Either I am sitting or I am standing.
Premise 2: I am not sitting.
Conclusion: Therefore, I am standing.

is a deductively valid argument because if the premises are true, then the conclusion in this case *must* follow – there is no other alternative. But consider now this contrasting case:

Premise 1: Coffee from my local coffee shop has not killed me in the past.
Premise 2: This coffee I am drinking now is from my local coffee shop.
Conclusion: This coffee I am drinking now will not kill me.

While this inference is certainly a reasonable one, in that the premises provide good grounds for believing the truth of the conclusion, it should be clear that they do not *guarantee* the truth of the conclusion. Thus, we have two contrasting types of reasonable inferences – reasoning from premises that guarantee the truth of the conclusion and reasoning from premises that support, but do not guarantee, the truth of the conclusion. All scientific knowledge is derived from this latter type of reasoning, which means that, even *in principle*, scientific reasoning does not ever give us 100% certainty, because it is not deductive in nature. But, of course, this does not mean that scientific reasoning is unreliable (as history shows us, quite the contrary is the case!) or that scientific uncertainty is inherently controversial. Instead, it simply means that we need to be aware of the fact that scientific reasoning always yields results with some level of uncertainty and that this should be openly acknowledged and communicated by scientists, to the public and to stakeholders generally.

Scientific expertise and moral authority

Once we are able to identify who counts as a scientific expert (keeping in mind that no expert is ever infallible and that scientific results are never 100% certain) then we can be reasonably assured that they will be able to inform us about what is the case, given some domain or some domain-specific set of questions. In other words, we can be reasonably certain that they will provide us with accurate information. However, scientific experts are not able, simply in virtue of being experts, to tell us *what to do* with the information that they provide. By now, this point will hopefully be familiar. But in case it is not: the reason for this is because moral action lies outside of the domain of science, by design (as we saw in Chapter 1), and therefore must always be supplemented with and informed by extra-scientific information and/or values. This is not to say that these sorts of extra-scientific values do not factor in to the methodology of science in the first place; they certainly do. Instead, the point is that:

> Science cannot make value judgments. Science does not determine policy. Policy is a human endeavor that combines science with values and priorities. In other words, science can [for example] help quantify the increased risk (or lack thereof) of school reopening on SARS-Cov-2 spread, and help quantify the educational losses from continued closure, but science cannot tell you whether to open or close schools. Scientists have no special ability to speak about values on behalf of all citizens.
> (Prasad 2020)

This is not to say that science itself, even before it is applied, either is, or should be, value free. On the contrary (Lekka-Kowalik 2010), as we saw in Chapter 4, "in the past few decades, philosophers of science have shown that even good science requires non-epistemic value judgments" (Schroeder 2019). For example, and perhaps surprisingly to some, even the outcomes of scientific measurements are more than just simple "facts.[14]" Take, for instance, the question of whether or not a medical diagnostic test is accurate. Answering this question is different than simply asking whether or not a given tape measure is accurate. In the case of a tape measure, all we need to do is compare the tape measure with some reference standard, such as the standard meter in Paris, in order to determine whether or not it is accurate. But to determine whether or not a given diagnostic test is accurate, we need to compare the measurement capabilities of the test with that of a reference standard and we also need to understand how to *interpret* the quantities that the test measures. For instance, to determine whether or not a given serum assay for cholesterol is accurate in diagnosing hypercholesterolemia, we need to know not only if the test accurately measures the amount of cholesterol in the blood but also whether or not to count whatever level it measures as 'high,' 'low,' or 'normal,' and this act of interpretation requires making non-scientific judgments (Kennedy 2015a).

Further, this intertwining of fact and value in science has practical application for citizens in a liberal democracy: it means that we "can and should embrace science, but we cannot follow it. It is up to us to make the hard choices" (Prasad 2020). So, while we can trust expert testimony to be helpful in forming reliable beliefs, more is needed in order to prescribe actions – particularly those actions which fall outside of the domain of science. This is because science cannot dictate policy, it can only inform it.

However, it should be noted that this *informing* of public policy is an incredibly important role for science, and should not be down-played. In fact, the collective actions that we take (or don't take) based on scientific findings have, in many cases, real and lasting impact on both local and global communities, as we will see in the discussion of the COVID-19 example that follows. Thus science plays a vital and indispensable role in policy formation, in that it can tell us what *is* the case, however, in order to *apply* science we must appeal to concepts that lie outside of it. As Cowley (2012) puts the point, "All scientists are answerable to a singular realm of discoverable facts. But the same facts may well have different moral significance for different individuals." This is an especially important point to understand in the context of a liberal democracy in which a multiplicity of values is often represented, and it should encourage us to adopt a pluralistic framework for the weighing of these values. The important point, though, is that the application of social and ethical concepts and values is always going to be necessary when using science to inform public policy. While this might be disconcerting to some who hope for an entirely dispassionate way to decide policy, in the end this is neither desirable nor possible.

Rather,

ethical values are inextricably intertwined with every determination of what counts as evidence in any given situation [...] While once touted as the ideal, "value-free" science is now recognized by many as a state that is both unreachable and undesirable. While those who once advocated for value-free science did so "because they understand values to be ideologically held and immune to rational evaluation," (Goldenberg 2016) many now recognize that values can in fact be amenable to revision given empirical evidence and thus that they can be *good* reasons, not just reasons, upon which to base decisions.

So although "there is no rule of logic that can help us decide whose interpretation of empirical experience is *the evidence*," (Shahar 1997) this should not be a cause for despair.

(Kennedy 2021)

Because there is no such thing as value-free science (in either methodology or application), many have argued that scientific experts who act in the role of policy advisors should make the extra-scientific concepts and values on which they base their judgments transparent to the public (Douglas 2009; Elliott 2017, 2019). Doing this, according to Douglas and Elliott, will help to maintain the integrity of science while also allowing for democratic accountability in policymaking. Of course, as Schroeder (2019) points out, making values explicit will increase public trust in science *only if* those values are *democratically decided* ones, rather than ones simply held personally by the scientists conducting the research (in which case these "values" would look, to the public, much more like preferences or biases, rather than anything more helpful). But given this caveat, making extra-scientific values explicit allows them to be publicly discussed and evaluated and this can, in turn, both help agency officials make better informed decisions and help to foster public trust in those decisions. It can also help to keep scientific experts accountable, and allow for the public to weigh in on the application of values to policy recommendations. This is important because:

> Scientists have no special expertise on these value judgments; in a democratic society ordinary people are entitled to criticize scientific theories for failing to incorporate certain values – or for incorporating bad values.
>
> (Anderson 2011)

However, there are also some potential dangers in making these extra-scientific values transparent. As Elliott (2021) argues, some of these dangers:

> are related to the time and money involved in making information transparent. Other dangers involve harm to companies, research participants, or natural resources that could arise by making confidential, private, or sensitive information available to others. Still other dangers are associated with the risks of confusing the recipients of information, equipping malicious actors with information they can use to harass scientists, or creating unjustified skepticism on the part of those who receive information.

However, proponents of transparency have argued that these kinds of difficulties can be alleviated by clear and careful communication (Elliott and Resnik 2014; Stanev, 2017; Pinto and Hicks 2019). What all of this means is that it is vital first, for experts to make their value judgments explicit, as well as available to public examination and second, that it is important for the public to ask the question of when, and how, we ought to supplement scientific information with other, non-scientific considerations and values when applying scientific results to public policy. This is because when "public policy claims to follow the science, citizens are asked not just to believe what they are told, but to follow expert recommendations" – and the only way to evaluate the rightness/wrongness or aptness/inaptness of a prescribed action is to appeal to human values. This, in turn, means that if "we are to ask the public to trust the recommendations of scientists, we must acknowledge that this is different from asking novices to accept facts" (Bennett 2020). When we are asking the public to accept a recommendation from an expert, we are asking those persons to "believe that the expert bases their recommendation on values that are held by the recipient of the recommendation," because recommendations do not simply "fall out" of the data alone. In other words, we when are asking the public to accept an expert's recommendation regarding an action, we are asking for a particular kind of trust in the expert – not simply trust that the expert is competent and sincere, but "also that their recommendations are in our interest" (Bennett 2020).

Consider an analogy from clinical medicine that helps to illustrate this point. Imagine that a physician (whom we might reasonably describe as a medical expert, assuming that they are both competent and sincere) advises their patient to have a certain surgery. To weigh whether or not to have the surgery, it is likely not enough for the patient to know that the physician is an expert (that is, that the physician is competent in their field and sincere). Instead, the patient will very likely also want to weigh whether or not, all things considered, the surgery is in *their own* personal best interest. And this is something that only the patient can decide (perhaps with the help of the physician's input), because it depends upon the patient's personal values and goals, etc.

This situation is similar to that of expert-informed policy decisions in the context of a liberal democracy. While we certainly do want to have the input of experts, we also want to avoid an erosion of democratic decision-making by allowing experts to make our decisions for us. In other words, some have argued regarding this concern that "there appears to be a tension between two demands – that public policies be empirically responsible and that they be democratically legitimate." The worry arises in part because a "decision that follows or is based on science does not entail a good decision or one that is better than what could be decided using something other than science" (Anderson 2011). Further, because scientific experts are not elected (as policymakers generally are) they are not held accountable to the population they inform or to the values that that population holds. Thus, policy (in a democracy) must be informed by democratically held values – because there simply is no such thing as either science or policy that is void of value judgment. And this is not a bad thing. In fact,

> we might be better to embrace the fact that human judgement is involved in any policy decision-making process than to rely on vague appeals to science. Making such judgement explicit when engaging the public could go far to engender trust, both in the political process and in science. We should also be vigilant to avoid letting the scientific evidence that does exist overwhelm other considerations that may warrant a change in policy or let one group of scientists dictate the discussion.
>
> (Mercuri 2020)

What this means is that appeals to science can and should be made when making policy decisions. However, it should also be recognized (and publicly admitted) that scientific data alone cannot dictate policy – human values, and in a democratic society, democratically decided ones – must also inform these decisions.

Example: COVID-19

As an example of how this might work in practice, let's consider the "lockdown" policies that were implemented by many governments during the crux of the COVID-19 pandemic. At the time when these policies were being implemented in the United States and elsewhere, during the early months of 2020, many decision-makers were appealing to the scientific data produced by the Imperial College London COVID response team's model (Ferguson et al. 2020) to inform their policy recommendations. This epidemiological model, which later became very controversial due to its strict recommendations, suggested that of the two available strategies for addressing the virus, prior to the availability of any vaccine, that suppression "which aims to reverse epidemic growth, reducing case numbers to low levels and maintaining that situation indefinitely" was "the preferred policy option" because it would result in fewer deaths than a strategy of mitigation, which was defined in the model as a focus on "slowing but not necessarily stopping epidemic spread" (Ferguson et al. 2020).

This model and its recommendations quickly became controversial because they advocated for a stringent "combination of social distancing of the entire population, home isolation of cases and household quarantine of their family members.... supplemented by school and university closures."

And although the creators of the model did recognize that such closures could "have negative impacts on health systems due to increased absenteeism," they did not acknowledge any other potentially negative impacts that such measures might cause, such as the inequitable distribution of financial burdens on differing populations globally, mental health repercussions for individuals experiencing forced isolation, or the inequitable impact of school closures on minority groups, etc. Instead, in their paper, the scientists who created the model argued that lockdown measures were preferable to allowing the greater number of virus deaths that mitigation strategies would entail, presumably because more deaths would be worse than the potential negative effects on health systems. It should be pointed out that this was not an *unreasonable* conclusion, especially given the fact that at the time when the response team's report was released, there had been 1,426,102 confirmed cases of COVID-19 and 81,865 deaths due to the virus. And that further, just two weeks after its release, and while its recommendations were still being debated by policymakers, the Prime Minister of the United Kingdom was in an intensive care unit of an NHS hospital, fighting a severe case of the virus. In other words, many people were sick, many were dying, and it was becoming increasingly apparent that the virus was affecting people of all socio-economic classes in all parts of the world. Still, even at that time, some were asking: are things really that bad? Is the whole world overreacting? Does the severity of the pandemic *actually* merit the suppression measures being suggested? These, of course, were, and remain, not only difficult questions to answer, but questions that no scientific model, on its own, can provide answers to, precisely because all of these questions involve social, ethical, and economic considerations in addition to scientific ones. Thus while those who agreed with the model recommendations argued that not following them was tantamount to a "rejection of science," they weren't really right about this: the model made predictions (some of which turned out to be accurate, while others did not) based on the best available epidemiologic data at the time. But this alone was not enough, because it could not be, to straightforwardly dictate policy decisions. In other words, those who objected to the model's recommendations were on to something when they accused these epidemiologists of overstepping their bounds. To be clear, this is not to say that the recommendations of the Imperial College model were wrong – that is (still) up for debate. But it *is* to say, that this is a clear example of how science alone cannot tell us how to act. What science can do, and what it did in fact do in this case, is give us estimations and predictions – in this case about the number of deaths that would occur when implementing mitigation strategies vs. suppression strategies. But what it could not, and cannot, do is tell us how to weigh the costs vs. benefits of, for instance, stay at home orders. That is, it could not tell us how to weigh the considerations of the number of COVID-19 deaths vs. the negative health impacts of reduced income due to lost work, or health disparities exacerbated by social isolation. As previously mentioned, on these issues, many people reasonably disagree. For instance, some (such as West 2022) have argued that the costs of lockdown fall primarily on the global poor, and that the benefits accrue primarily to the global rich, while others argue that, even so, these costs are worth the benefits, because the benefits mean the saving of lives, which trumps all other considerations.

But even aside from these competing ethical concerns there is also another issue to consider here, and that is question of what sort of scientific data counts as accurate enough to merit action. This question is closely connected to the question of what counts as scientific evidence in the first place, which we discussed in detail in Chapter 2. But here, we can further extend that line of analysis in the following way. Suppose, for instance, that the outputs of the epidemiological model created by the Imperial College team did in fact meet the standards of scientific evidence – that there were no major mistakes in the modeling and that the model employed the best information available at the time. Yet, we still must take into consideration the fact that not all scientific evidence is on a par, but that some can be considered to be stronger than others. The question that then arises when we want to use scientific data, or evidence, as input into our policy decision-making is that of *how strong* it must be in order to merit action. This, of course, will depend on what would happen if action were *not* taken. But this too, is often an estimate. In the case of this particular model, the scientists recommended the actions that they believed would lead to the fewest number of viral deaths, and thereby implicitly argued against inaction, since that would inevitably result in a higher number of deaths. However, not everyone agreed with these recommendations. Some disagreed because they thought that there could be other outcomes from lockdown measures that were just as bad as (or worse than) viral deaths (Simon et al. 2021). Others disagreed because they argued that no actions should be taken based a model that did not reach a certain standard of evidence that was "morally required" (Winsberg et al. 2020), regardless of the outcomes of those actions. In other words, the argument was that, this particular model as not good enough to merit restrictions of any kind – even if those restrictions did end up reducing the number of deaths. That is, Winsberg et al. argued that the ends do not justify the means and that, in particular, this model did not reach the morally required epistemic threshold necessary to justify infringement upon basic human liberties. They wrote that:

> decision-makers "have moral duties to collect good evidence, [and to] reason carefully about that evidence," particularly when this evidence will be used in high-stakes decisions and that, further, "meeting

the epistemic duty means relying on good information—not the best information available, but good information, period".

Of course, this line of arguing also begs the question of what counts as a "high-stake" decision, as well as what the morally required epistemic threshold should be – in other words, how high is "high enough" to merit action, particularly when inaction, as in this case, was, in effect, a kind of action, with potentially devastating consequences? Douglas (2016) articulates this question in the following way:

> A scientist always needs to decide, precisely [...] whether the available evidence is enough for the claim at issue. This is a gap that can never be *filled*, but only stepped across. The scientist must decide whether stepping across the gap is acceptable. The scientist can narrow the gap further with probability statements or error bars to hedge the claim, but the gap is never eliminated. How is a scientist to decide that the available evidence is enough? That the gap is worth stepping across?

To answer these questions, Douglas further argues, the scientist must appeal to values, which can be thought of as further forms of evidence, that necessarily lie outside of science. Thus, we can see from the discussion of this example that science alone is not enough to tell us what to do when it comes to policy decisions: it can give us data, but it cannot tell us how to act. For that, we must appeal to non-scientific concepts, values, and ideas. In the following chapter, we will see how this can be done in actual practice.

Conclusion

The question of scientific expertise, and, in particular, that of who counts as an "expert" is closely connected to the issue of science communication, which we discussed in detail in Chapter 4. Here in this chapter, I have attempted to show that determining what it means to be an "expert," and, in particular, a scientific expert, in the public arena, matters in another way, because there is a relationship between scientific expertise and what is known as epistemic authority. What this means is that scientific experts, even though they are not infallible, nor are they immune to the constraints of scientific uncertainty, are able, in virtue of their expertise, to convey information that allows non-experts, including members of the general public as well as other stakeholders, to improve their epistemic understanding in a given domain, and thereby to inform public policy decisions in relevant ways. I have also proposed that a scientific expert is someone who is both competent in their field *and* is trustworthy, or sincere, regarding the information they convey. This means that any "expert" who is not sincere is not, on this view, an expert at all.

Further, we have also seen via an examination of the 2020 Imperial College COVID-19 model, that although we ought to trust scientific experts because we can learn from them what *is* the case, and thereby increase our knowledge base by consulting them, scientific expertise alone is not enough to tell us how we *ought* to act. To know how to act on scientific information – even accurate information that is derived from experts – we must appeal to social, political, and moral values: there is no way around this (nor would we want there to be), and thus there is no such thing as simply "following the science." Science is a method, albeit a reliable one, but it is neither a tour guide, nor a simple prescription for action. To decide how to act we must appeal to human values, and these necessarily lie outside of the domain of science.

Notes

1 The next chapter will further address the question of whether or not there exists a sharp distinction between "the experts" and "the public," as well the question of as whether or not (or to what extent) this distinction is helpful in the realm of policy making, particularly in a democratic context.
2 https://www.independent.co.uk/news/world/americas/us-politics/fauci-interview-today-science-covid-b1862899.html.
3 Although I use the term "expert" in the singular here, it should made be clear that experts don't have to be individuals, but can also be panels or commissions, etc., comprised of more than one individual.
4 He leaves open the question (as do I) of exactly how many more true beliefs an expert must have as compared to a non-expert. However, it is clear that we don't want to set the bar too high (even experts get things wrong some of the time) nor too low (51% true beliefs and 49% false, for example, probably isn't what we want in an expert).
5 This added requirement for expertise is, I think, similar to the justification requirement appealed to in the justified-true belief (JTB) account of knowledge in contemporary epistemology, where a belief counts as knowledge only if it is both true *and* justified, rather than true simply by luck, or by some other means.
6 While experts are required to be "epistemically trustworthy," we might say that non-experts, on the other hand, have a duty to be "epistemically vigilant" (Sperber et al. 2010), which means that they must not put blind trust, but only *warranted* trust, in any purported or potential expert. In other words, they should have reasons for trusting the experts in question.
7 Some might want to call an insincere or unreliable "expert" a "pseudo-expert." I prefer to simply say that such a person (or panel) would not count as an expert at all.
8 Hereafter I will simply use the term "expert" to refer to an expert in the public domain.
9 Anderson (2011), for example, suggests that we can construct a hierarchy of scientific expertise, from lowest to highest, in the following way:

a Laypersons.
b People with a BS degree, or a BA science major.

c PhD scientists outside the field of inquiry.
d PhD scientists trained in the field.
e Scientists who are research-active in the field.
f Scientists whose current research is widely recognized by other experts in the field,
g Scientists who are leaders in the field – who have taken leading roles in advancing theories that have won scientific consensus or opened up major new lines of research, or in developing instruments and methods that have become standard practice.

10 Although it is interesting (and perhaps also frustrating) that non-experts have to appeal to experts to determine who counts as an expert, or at least to determine the competency component of expertise.

11 Of course, television interviews can sometimes portray misleading information – by focusing on soundbites, or by asking "leading questions" of interviewees. My aim here is not to make a determination of whether or not Dr. Anthony Fauci, in particular, is or is not trustworthy (or is or is not an expert), but rather to give an example of a purported expert relaying inaccurate information (whether intentionally or not) in a public setting.

12 Certainly, all experts (and indeed all humans) at times make mistakes, so we should not rule out every human who makes a mistake as a potential expert. Science itself, as we have seen in previous chapters, is not infallible, so it would be unreasonable to expect scientific experts to be either. However, anyone who regularly conveys misinformation (either knowingly or not) should not be understood to be an expert, and this is because experts must be, among other things, reliable.

13 This also means that paternalistic "experts" insofar as they mislead the public "for their own good" would not, on this definition, count as experts because experts, must be, on the balance, truthful in their public statements and communications.

14 Here I mean to distinguish between "fact" and "truth." Facts, in order to be such, must certainly be true, but that doesn't mean, of course, that their formation must be devoid of any/all values or opinion as input. In general, as previously discussed, I think that the fact vs. opinion distinction is completely unhelpful. I use the term "fact" here only because it is commonly used in the literature on science and public policy.

6 Science-informed public policies

[While] science has been very effective in bringing issues into the public arena, it has been quite ineffective at providing solutions.

(Jamieson 1996)

Introduction

One of the primary aims of the preceding chapters of this book has been to prepare us to understand how to apply the data that we derive from scientific models, observations, and experiments to our public policy decisions. To do this this effectively, there are several steps that must take place, and during the course of this chapter, we will examine each of them, in turn. In particular, we will take a closer look at the ways in which both scientific evidence as well as extra-scientific considerations play a role in each of these steps:

> Policy making, far from being a sphere in which science can be neatly separated from politics, is a sphere in which they necessarily come together.
>
> (Jasanoff 1990)

Further, as we will see, every application of science to policy begins with the identification of a specific problem – because, of course, if there is no problem (or potential problem) then there is no reason for a policy to be enacted. However, not all potential problems merit policy intervention, and further, of the ones that do, not all require scientific input, although most of them do:

> Just about every public policy issue has a scientific component, whether it's hidden or obvious; but the facts and insights science brings to deliberations can often go unaddressed.[1]

DOI: 10.4324/9781003311072-6

The issues that we will be interested in, as the example that we will examine in this chapter will help to illustrate, are the type that require an intervention, and, in particular, an intervention that is adequately informed by the relevant scientific data in addition to certain extra-scientific considerations. In other words, this example will help to illustrate the points that:

> Policy is made in many settings. It evolves from a many faceted social process involving multiple actors engaged in assembling, interpreting, and debating what evidence is relevant to the policy choice at hand, and then, perhaps, using that evidence to claim that a particular policy choice is better than its alternatives.
>
> (Cairney and Oliver 2017)

To begin, we will first need to know how to identify issues that are worthy of public policy interventions. In general, these are of the sort that have a significant impact on one or more populations of people and/or that impact society in a more general way. As we saw in Chapter 3, the problem of secondhand smoke inhalation is one example of this, and, as we saw Chapter 5, the COVID-19 pandemic is another. These problems were determined to merit policy interventions because they were seen as issues that posed significant population-level health threats. Further, both of these issues had both scientific and ethical components to them. In the case of secondhand smoke, the available scientific data showed that inhaling it caused certain negative health effects and this information, coupled with the normative judgment that this ought to be addressed, prompted the subsequent enaction of public smoking bans in this country. In the case of COVID-19, the situation was similar: science identified the virus, and the fact that it was spreading, but it was only with the additional (extra-scientific) value judgment that this was a bad thing that should be changed, that the problem was determined to be one that merited intervention.

As we already saw with these examples from previous chapters, and as we will see in more detail in this one, the next step, after the identification of a policy-worthy problem, is to propose a potential solution, and then to communicate the proposed plan to the relevant non-scientist stakeholders, including both policymakers and the general public. In doing this, it is of particular importance for these parties to agree on the justification for the proposed intervention. This, in turn, requires that scientists and other experts both acknowledge any uncertainty in their data and/or predictions as well as any and all appeals that they make to extra-scientific values, so that other stakeholders can genuinely weigh in on the decision-making process. And finally, the last two steps in any application of data to public policy are to implement the proposed solution and then to follow up with research to find out if it was effective in achieving its aims. To see how each of these steps work in actual practice we will now take a look at an example.

Example: laws governing the sale, prescription, and use of addictive substances in the United States

Substance addiction is a serious and widespread problem that affects millions of individuals worldwide. According to data from the United Nations:

> about 5.5 per cent of the population aged between 15 and 64 years have used drugs at least once in the past year, while 36.3 million people, or 13 per cent of the total number of persons who use drugs, suffer from drug use disorders.
>
> Globally, over 11 million people are estimated to inject drugs, half of whom are living with Hepatitis C. Opioids continue to account for the largest burden of disease attributed to drug use.[2]

Further, when it comes to the United States specifically, according to the National Institutes of Health (NIH), the Center for Disease Control (CDC) and the National Drug Intelligence Center:

> [The] use and misuse of alcohol, nicotine, and illicit drugs, and misuse of prescription drugs cost Americans more than $700 billion a year in increased health care costs, crime, and lost productivity. Every year, illicit and prescription drug overdoses cause tens of thousands of deaths (nearly 70,000 in 2018), alcohol contributes to the death of more than 90,000 Americans, while tobacco is linked to an estimated 480,000 deaths per year.[3]

In addition to this:

> Among Americans aged 12 years and older, 37.309 million were current illegal drug users (used within the last 30 days) as of 2020, 13.5% of Americans 12 and over used illegal drugs in the last month, 9.277 million or 21.4% of people 12 and over have used illegal drugs or misused prescription drugs within the last year, 138.543 million or 50.0% of people aged 12 and over have illicitly used drugs in their lifetime, 28.320 million or 20.4% have an alcohol use disorder, and 24.7% of those with drug disorders have an opioid disorder; this includes prescription pain relievers or "pain killers" and heroin.

Substance addiction is a problem because it often leads to significant health, relational, and social issues for addicted individuals, their loved ones, and their communities at large. For example, methamphetamine use can cause severe dental problems, opioid use can lead to overdose and death, and intravenous drug use can increase the risk of contracting serious infections such as HIV, hepatitis C, endocarditis, and cellulitis.[4]

But the negative consequences of substance addiction are not limited to those who use the substances themselves – in addition to the serious and detrimental consequences that can occur for the addicted individual, the impact of substance addiction on family and other personal relationships is also significant and costly:

> Each family and each family member is uniquely affected by the individual using substances including but not limited to having unmet developmental needs, impaired attachment, economic hardship, legal problems, emotional distress, and sometimes violence being perpetrated against him or her. For children there is also an increased risk of developing an SUD themselves.
>
> (Lander et al. 2013)

Compounding these individual and relational problems is the additional fact that, according to the National Institutes of Health:

> The substantial prison population in the United States is strongly connected to drug-related offenses. While the exact rates of inmates with substance use disorders (SUDs) is difficult to measure, some research shows that an estimated 65% percent of the United States prison population has an active SUD. Another 20% percent did not meet the official criteria for an SUD, but were under the influence of drugs or alcohol at the time of their crime.
>
> (Center on Addiction 2010)

Thus we can see from even a cursory examination of these statistics that the consequences of drug addiction are serious, wide-ranging, and have measurable and negative effects on the health of individuals, on their productivity, on their interpersonal relationships, as well as on crime rates, and on social welfare in society more generally (National Institute on Drug Abuse). Because substance addiction is recognized as such a significant, widespread, and growing problem, a variety of laws and other regulations have been implemented in the United States in recent years, which aim to govern the use, sale, prescription, and/or possession of addictive substances, in an effort to address the problem. Here in what follows we will take a closer look at some of these laws, at the reasons behind implementing them, and at the question of whether or not they are, on the balance, effective in achieving their aims. Further, we will use this specific example of the application of scientific data, combined with extra-scientific value judgments, to public policy and law in order to illustrate some of the more general considerations that arise when we put science into practice in this way.

What is addiction?

Before addressing a problem such as addiction, it is necessary to have at least a rough understanding of what the condition is. While this might seem to be obvious, in the case of addiction, it is far from a simple task, as is evident from the ongoing and considerable disagreement amongst those who work on the topic (Wakefield 2019). Indeed, as Morse (2012) puts it: "virtually every statement that can be made about drugs and addiction, whether it is factual or normative, is contestable" (p. 262). There are many potential reasons for this, one of which is that there are a variety of kinds of addiction, not all of which involve substances, such as sex addiction, gambling, video game addiction, etc. And this further complicates the already difficult matter of defining what exactly addiction is. However, for our purposes in this chapter, we will focus solely on a discussion of substance addiction, leaving questions about other sorts of addictions aside, which will make the task somewhat easier, yet still significantly complex.

In the specific case of substance abuse addiction, there is considerable controversy over whether or not this condition should be considered to be a disease, a choice, or some combination of the two, as well as over whether or not the condition is more likely to be caused by genetics or by environmental factors:

> Recent neuroscientific developments fuel the view that addiction can be classified as a brain disease, whereas a different body of scholars disagrees by claiming that addictive behaviour is a choice.
>
> (Goldberg 2020)

On the one hand, the NIH, as well as the National Institute on Drug Abuse (NIDA), for example, weigh in strongly on the disease side of the debate:

> As a result of scientific research, we know that addiction is a medical disorder that affects the brain and changes behavior. We have identified many of the biological and environmental risk factors and are beginning to search for the genetic variations that contribute to the development and progression of the disorder. Scientists use this knowledge to develop effective prevention and treatment approaches that reduce the toll drug use takes on individuals, families, and communities.[5]

In other words, according to the NIH and the NIDA, there is no moral failing involved in substance addiction, as it is not a matter of willpower or choice, but instead is a disease of the brain. However, others disagree (such as Pickard 2019), arguing that the situation is more complex than this, for several different reasons. One of these reasons is that not everything that negatively affects the brain is something that should count as a disease

(Wakefield 2019). For example, we might not want to classify a brain injury due to a car accident as such. Another reason is that the concept of disease is itself value-laden (Morse 2012), that is, in order to determine whether or not some biological condition is a disease requires us to invoke more than just science, because it requires us to answer, amongst other things, the question: Are diseases determined by the natural world or are they human-imposed? Very generally, we can classify those who answer the former as "realists" about disease. Realists argue that diseases are things that are "out there" in the world, so to speak, and that can be defined scientifically. On the other hand, those who answer the latter are what are known as "constructivists." Constructivists hold the view that human interests, rather than biological malfunctions, explain disease. That is, according to constructivism, diseases cannot be identified independently of human values. Rather, on the constructivist view, to call a condition a disease is to make a judgment that someone in that condition is undergoing a specific kind of harm that we explain in terms of bodily processes. To put it simply, on this view, diseases don't "occur," but instead are constructed (Simon 2017).

However, to complicate matters further, before we can even begin to weigh in on one side or the other on the question of what it means to have a disease, we first need to determine the difference between a state of disease and a state of health to begin with. But this is not a simple task, nor is it one that can be done via appeal to science alone. For instance, some argue that health is just the absence of disease – that is, that a healthy human body is one that is in a state that is free from negative biological disruption. Yet here we can see that this view is not a purely biological one, since it appeals to the normative concept of "negative." Other definitions, such as the one proposed by the World Health Organization (WHO) which defines health as "a state of complete physical, mental and social well-being and not merely the absence of disease or infirmity,"[6] appeal to normative concepts as well. And although a complete discussion of the concept of health vs. disease is well beyond the scope of this chapter, we can see from even this brief overview that what it is to be healthy vs. what it is to be ill, is not always agreed upon, nor is it a question that can be decided by an appeal to science alone.

Still another reason for potential disagreement with the view put forth by the NIH and NIDA that addiction is a disease is that, even if we accept that it is, this does not mean that willpower plays no role in the condition. In fact, according to the American Medical Association and the American Society of Addiction Medicine, addiction is caused by a combination of behavioral, psychological, environmental, and biological factors, and in this way, is similar to both heart disease and type 2 diabetes:

> Choice does not determine whether something is a disease. Heart disease, diabetes and some forms of cancer involve personal choices like

diet, exercise, sun exposure, etc. A disease is what happens in the body as a result of those choices.[7]

Further and closely related to the question of whether or not addiction is a disease, is the question of what role genetics vs. environmental factors play in those who develop it. In general, it is considered to be the case that:

> Both genetic and environmental variables contribute to the initiation of use of addictive agents and to the transition from use to addiction. Addictions are moderately to highly heritable. Family, adoption, and twin studies reveal that an individual's risk tends to be proportional to the degree of genetic relationship to an addicted relative. Heritabilities of addictive disorders range from 0.39 for hallucinogens to 0.72 for cocaine.[8]

It is also generally agreed upon that, while there is no specific identifiable gene that directly causes addiction, that certain "genes of vulnerability increase the risk to ultimately develop an addiction" (Gorwood et al. 2019). Thus, both those who think that addiction is due to disease, and those who do not, believe that genetic tendencies play at least some role in the development of the condition.

But why is there controversy over whether or not addiction is a disease or a disorder in the first place? In other words, why does the answer to this question matter? It matters, because the way that we respond to this question has implications for both scientific research and for the public policies that potentially result from it. In particular, the question of whether or not addiction impairs self-control has important implications for both research and policy. Further, in addition to this, as Henden (2019) argues, impaired self-control in the context of addiction seems to be something that we all, at least prima facie, accept (even if we believe that addiction is caused by a disease of the brain), and thus this gives us a common-sense reason for believing that it plays an important role in the condition:

> One reason for thinking that addiction impairs self-control might be that it seems close to being a conceptual truth that a person who is actively addicted to a drug must have impaired self control over her consumption of that drug. Consider, for example, the suggestion that a certain nicotine addict who smokes 60 cigarettes per day is in "full control" of her smoking behavior. I think most people would be inclined to find this claim odd, indeed, even contradictory.

(p. 49)

Indeed, to press the point further, it seems fair to say that "most people-laypeople and specialists alike-regard addiction as causing or constituting a pathology of control over behaviour" (Levy 2019). And we should recognize that this is not necessarily a bad thing, as it can serve to give us insight into how to effectively treat and/or manage the condition.

In general, on both views, that is, on both the disease and the self-control models, substance addiction is thought to be a complex condition that includes the following characteristics:

> craving (preoccupation/anticipation), binge/intoxication, and withdrawal/negative affect. Impulsivity and positive reinforcement often dominate the first stages, driving the motivation for drug seeking, and compulsivity and negative reinforcement dominate the terminal stages of the addiction cycle. Addictive drugs induce adaptive changes in gene expression in brain reward regions, including the striatum, representing a mechanism for tolerance and habit formation with craving and negative affect that persist long after consumption ceases. These neuroadaptive changes are key elements in relapse.[9]

Thus we can see that addiction is a complex condition that is not easy to define, and yet it is one that still requires at least a working definition, for (Levy 2019, p. 59). However, in spite of the complicated nature of addiction, as well as the significant difficulties in defining it, it is interesting to note that many of the scientists who work in this area of research write and speak as if what addiction amounts to is a clear and settled matter. But perhaps this should not surprise us, as the primary role of a scientist is one of investigation, rather than of creating definitions or engaging in conceptual analysis. Yet even so, in many instances, and certainly in the case of addiction research, the questions that scientists investigate require, before they can even get off the ground, so to speak, at least working definitions of sometimes very complex concepts. To be more specific, this means that in order to answer the question of whether or not a given substance is addictive, we have to have at least some idea of what addiction amounts to. And we need to know this, not only for research purposes, but also for purposes of application. By analogy:

> The United States once passed a law that provided federal funds for public transportation systems in any city in the United States. Many communities applied for the funds, but which of these communities were cities and, hence, eligible for funds? The government could not answer this question by citing common usage, because common usage is too vague. The government also did not want to decide arbitrarily, since that would be unfair and impossible to justify to communities that were excluded. What they did was to estimate how many communities would apply and for how much funding. Then they could calculate how far the available

funds could go. They defined "city" so that their available funds could help as many people as possible.

(Sinott-Armstrong and Summers 2019)

This is an example of what is known as a pragmatic definition, and such definitions are also sometimes used as working definitions for the purposes of scientific research. And while pragmatic definitions are unsatisfying for some, because they do not answer the conceptual question of what something *is,* in many cases they are necessary, at least temporarily, in order for research and application to proceed. These sorts of definitions give researchers and policymakers something to work with, while they proceed with studying not yet well-understood entities, concepts, or processes, such as addiction.

While the question, "what is addiction?" is one that is important for both research and application purposes, in that these activities cannot proceed without at least some sort of answer to it, it is only *partly* one of science, and this can sometimes complicate matters, particularly when it comes to the development of public policies, as we will see in more detail in the example that follows. However, before we get to those details, and in order to begin the discussion, it is helpful to first take a look at some of the ways that addiction is currently defined by various international and national health research organizations. First, according to the NIH:

Addiction is defined as a chronic, relapsing disorder characterized by *compulsive* drug seeking and use despite *adverse* consequences.

In addition to this, the National Health Service (NHS) in the United Kingdom says, similarly, that:

Addiction is defined as *not having control* over doing, taking or using something to the point where it could be *harmful* to you.

However, the World Health Organization (WHO) defines addiction, in a slightly more complex way as a:

treatable, chronic medical disease involving complex interactions among brain circuits, genetics, the environment, and an individual's life experiences. People with addiction use substances or engage in behaviors that become compulsive and often continue despite harmful consequences.[10]

One thing that we can immediately see from an examination of each of these various, yet closely related definitions, is that there is a detailed conceptual analysis that is required in order to make sense of them – they are not necessarily straightforward to understand, but rather require a significant amount of interpretation. The first thing to notice is that each of these definitions is comprised of two prongs, or parts, so to speak. The first part of each definition refers to "compulsion" or an impulse over which one does not have control: a compulsion is something that is thought to "strip a person of all choice and power to do otherwise" (Pickard 2019). Another way to put this is that a compulsion is something that is completely irresistible. However, even here, right from the start, in examining just the first part of each definition, we can recognize a potential problem: addictive impulses cannot literally be irresistible or else recovery would never be possible – and we know that, at least in some cases, it certainly is. And this, it should be noted, is true even if addiction "involves functional changes to brain circuits involved in reward, stress, and self-control" (Goldstein and Volkow 2011).

Further:

> there is increasing evidence that addicts are not compelled to use: They are responsive to incentives, suggesting that the desire is not irresistible.
> (Pickard 2019)

The National Institute on Drug Addiction, even given their acceptance of the disease model of addiction, recognizes this as well. In their (2014) guidelines for treating addiction they write that: "Rewards and sanctions are most likely to have the desired effect when they are perceived as fair and when they swiftly follow the targeted behavior." Clearly, these measures would not be recommended if addicts were not able to respond to them.

And what this means, in turn, is that:

> Addictions are not necessarily compulsive […] if enough is at stake in someone's life, then it might not be unreasonable to expect, or indeed demand, that she (genuinely seek help to) overcome the problem.
> (Watson 2010)

However, if an addiction is not something that involves a true compulsion, then what exactly does it involve? Certainly, it must involve, at the least, an appetite, or kind of desire, for something. And yet, this still isn't enough for our definitional purposes because there are many things that we have appetites for (such as healthy food, clean water, or even relaxing TV time, that we would probably not want to count as addictions. This is where the second "prong" of the proposed definitions becomes helpful. If we define an addiction, not as something that is either a) literally irresistible or b) merely desirable, but instead as an appetite for something that is *also* in some way harmful to the person – physically, emotionally or otherwise, then we might be getting closer to the mark. In other words, as Watson (2010) puts it, we might want to say that an addiction is an appetite that, typically, is caused and sustained by the regular ingestion or use of certain substances *and* that causes some kind of harm to the person who ingests or uses them. It seems, then, that this leads us toward a behavioral/normative definition of what an addiction is. The definition is behavioral, because it describes the habits of those who are addicted. But it is also normative, because we don't classify habitual behavior as addictive unless it is also in some way harmful to the person.

This then leaves us with (the considerable task!) of defining what counts as harm, which in the case of addiction research is usually parsed as "negative consequences,"[11] in combination with deteriorating reward, due to the substance that is being used. However, we must be clear in our recognition that what counts as negative, or harmful, is not, as we have already seen in detail with other examples in preceding chapters, a question that science alone, without added input from other sources, can answer for us. To review, this is because the question of whether or not something is harmful, or has negative consequences, is a normative, and not a scientific, question. And this is because, as we learned from the discussion in Chapter 2, normative questions fall outside of the domain of scientific inquiry, a domain which is limited, by design, to the realm of empirical enquiry, which is that of space, time, energy and matter. If an entity or concept lies outside of this domain, then that means that it is not possible to devise any experiment, observation, or test that would allow us to learn anything about what that entity or concept is, or is like. This is the case with the concept of harm – we cannot devise an experiment which will tell us whether or not some substance, activity, or state of affairs is "harmful" or not. To see why not, consider the following example from clinical medicine as an illustration. Blood tests for things such as serum glucose level can, if they are reliable ones, provide us with an accurate number value that we can trust, and that we can potentially use for various applications. However, tests like this cannot, alone, tell us what the numbers they provide us with *mean*. This is something that requires added interpretation. For instance, to determine whether or not a given glucose level is good or bad, or even "high," "low," or "normal" we have to incorporate extra-scientific considerations into our analysis. To put it another way, science, often coupled with technology, can provide us with accurate data, but it cannot tell us what that data means, or how to apply it. These things require philosophical analysis, which means that all scientists must engage in philosophical enquiry, at least to this extent. To elaborate further, imagine that there is a patient who comes into the clinic complaining of severe fatigue. A serum comprehensive metabolic panel reveals a glucose level of 270. The clinician will then need to ask what this number means. The first thing she will probably do when attempting to answer this question is compare this value to the provided lab range. Doing this will reveal that this number is "high," as compared to the serum glucose level in a normal population. However, this first step of interpretation is still not enough to tell us whether or not the number 270 is, in this instance, good, bad or neutral. Why not? Because of the fact that just because this number deviates from a statistical norm, this does not necessarily mean that it is bad (or good). Take the measure of an intelligence quotient (IQ) as a parallel example. In this case, a high IQ, which by definition deviates from the statistical norm is, all things considered, considered to be a *good* thing. As another example, notice that the opposite holds for average body mass index (BMI) values in the United States. In this case, having a lower than average (although, of course, not too low) BMI compared to the average American adult is, in many cases, considered to be a good thing rather than a bad one. From this we can see that data alone does not give us value judgments.

But why, exactly, cannot the numerical data provided by scientific inquiry tell us whether or not something is good or bad? Philosophers have traditionally answered this question via an appeal to what is often known as "Hume's law," after the 18th-century philosopher who presumably first expressed it. Hume's law says that knowing what *is* the case (as science aims to tell us) cannot tell us what *ought to be* the case. Instead, only extra-scientific, philosophical inquiry can do that. And if this is right, then the data, numbers, and statistical analysis that scientific inquiry provides us with cannot tell us whether a given value is good or bad or neutral.

But if scientific data cannot do this, then what, exactly, can? Here is where extra-scientific considerations enter. Before we get in to those details in the context of the example that follows, however, it is worthwhile to first address a commonly stated objection to the view that science is powerless to answer normative questions. Likely, some of you who are reading this already have this objection in mind: you might be thinking, well, Hume lived way back in the 1700s, and things were clearly very different then! In particular, look at just how much science has progressed since that time! Isn't it possible that quantum physics or neuroscience, for example, is now equipped to tell us what is right and what is wrong, what is good and bad? (Caruso and Flannigan 2018). After all, since science is so successful in telling us about the natural world, why can it not also be successfully used to tell us what is good and what is not? In answering this objection, it is first of all important to notice that implicit in these questions is the considerable epistemic assumption that there is no method of knowledge-gathering that is superior to science (Hunter and Nedelisky 2018). Or, to put it another way, the substance of this objection lies in the implicit claim that there is no domain-limit on the kinds of questions that science can answer. But this is itself a substantive, and highly controversial philosophical claim. It is also one which I have consistently argued against over the course of this book. And although a full examination of the debate concerning what questions science can and cannot answer falls outside of the scope of this book, we do have strong reasons in support of the claim that the domain of science is limited to the empirical world of space, time, energy, and matter. If we then supplement this claim with a further one, that moral notions such as good and bad lie outside of this domain, then we necessarily arrive at the conclusion that science cannot tell us anything (on its own) about these concepts. Thus the short answer to the question of why science cannot tell us what is, fundamentally, at the bottom level, good or bad, is because these normative concepts are not empirically detectable and fall outside of the domain of scientific inquiry – again, by design. That is, because scientific inquiry presumes methodological naturalism, which, as we saw in Chapter 2 is the philosophical assumption that the *answers to scientific questions can only be given in terms of natural causes*. What this means, in turn, is that while science often does enable us to discover things that are useful or pragmatically advantageous for various purposes, and can even, in some cases, tell us how to reach whatever states we count as "good," or to avoid those we count as "bad," it cannot tell us what constitutes the good or the bad at the fundamental level in the first place. And again, this is not an issue of technology, nor is it one that will change as science continues to progress. Rather it is a problem of philosophical foundations: "good" and "bad" are not detectable things in the way that supernova or neurons are, and they never will be no matter how much our technology or scientific methods advance. Thus scientific inquiry, before it can even begin to investigate ways in which to make our lives better, or at least prevent them from becoming worse, must start with the philosophical assumptions of what counts as good and bad in the first place. At its beginnings, then, science is nearly "indistinguishable from philosophy" (Hunter and Nedelisky 2018). What this means is that in order to be good scientists (and eventually good policymakers) we must first engage in philosophical inquiry and conceptual analysis. Once we do this, and settle upon a working definition of what counts as bad, or harmful, then and only then, will we be equipped to determine which problems ought to be addressed by our policies. And here, as we will see, science certainly does play a role.

The solution

As we have already seen, in order for a given problem to be considered worthy of a public policy intervention, it must not be merely a scientific problem, but must also have a moral component, and, in particular, one that affects one or more populations of people that could potentially be benefited via a policy intervention that addresses it. In the specific case of substance addiction, science can describe the condition for us, and help us to define it, and, coupled with this, philosophical analysis can tell us that the condition is bad for both the communities and the individuals it affects. From this, we can then conclude that substance abuse is a public health problem that merits policy intervention measures.

But how do we go about enacting these interventions? As we saw in Chapter 4, the first step in this process is to communicate with potential stakeholders regarding potential solutions to the problem in question. In this case, as with many other public policies, the proposed intervention into the problem takes the form of legislation. But before any legislation can be implemented, it must be justified to the public, and this requires an appeal to extra-scientific values. To begin, in our specific example:

> The general justification for the legal regulation of addictive substances is best summed up in the preamble to the Federal Drug Abuse Prevention and Control Act:
> The Congress makes the following findings and declarations:
>
> 1 Many of the drugs included within this subchapter have a useful and legitimate medical purpose and are necessary to maintain the health and general welfare of the American people.
> 2 The illegal importation, manufacture, distribution, and possession and improper use of controlled substances have a substantial and detrimental effect on the health and general welfare of the American people.
>
> (Morse 2012)

Thus, the idea is that a) drugs, or other potentially addictive substances are not simply, in and of themselves, either entirely "good," or "bad," yet at the same time, they b) carry the substantial risk of imposing "detrimental effects" on the health and general welfare of communities, which in this case, is the American public. Because of this, legislation regulating them is deemed to be warranted. But why, exactly, is it warranted? To answer this question, legal theorists generally appeal to what is known as the "harm principle" which is the idea that individuals in a society do not have unrestricted autonomy, but rather, that the state is justified in intervening to block the actions of individuals when those actions cause harm. In the case of harm to others (such as rape, theft, etc.) appeals to the harm principle in the context of legislation are generally unproblematic. However, in cases in which the law in question is intended to protect someone from themselves, the justification is trickier, and many times contested. However, in the case of substance addiction, the laws that are passed to address it are not, in general, justified, via any appeal to paternalistic motives, but rather via pointing out that the actions of addicted individuals do not only affect the addicted persons themselves, but also the individuals and communities around them. Thus, the idea is that the state is justified in enforcing such laws in the interest of protecting its citizens from harm.

What form, then, do these laws take? In particular, in the United States, the laws governing addictive substances can be divided into three general categories.

In the first category are laws that limit access to addictive substances. In the second category are laws that criminalize the use and/or possession of these substances. Of note, it is often the case that the same piece of legislation, such as the Controlled Substances Act (CSA), will address both of these categories at once:

> The Controlled Substances Act (CSA) – Title II of the Comprehensive Drug Abuse Prevention and Control Act of 1970 – is the federal U.S. drug policy under which the manufacture, importation, possession, use and distribution of certain narcotics, stimulants, depressants, hallucinogens, anabolic steroids, and other chemicals is regulated.[12]

In the third category of laws governing addictive substances are those legislative measures that are aimed at facilitating the treatment or recovery of addicted individuals. For example, H.R. 6, the Substance Use-Disorder Prevention that Promotes Opioid Recovery and Treatment (SUPPORT) for Patients and Communities Act of 2018, was enacted in order to address the opioid overdose epidemic:

> The legislation includes provisions to strengthen the behavioral health workforce through increasing addiction medicine education; standardize the delivery of addiction medicine; expand access to high-quality, evidence-based care; and cover addiction medicine in a way that facilitates the delivery of coordinated and comprehensive treatment.[13]

Another example of this third type of law is the CARA act:

> The Comprehensive Addiction and Recovery Act (CARA) of 2016 authorizes over $181 million each year (must be appropriated each year) to respond to the epidemic of opioid abuse, and is intended to greatly increase both prevention programs and the availability of treatment programs. CARA launched an evidence-based opioid and heroin treatment and interventions program; strengthened prescription drug monitoring programs to help states monitor and track prescription drug diversion and to help at-risk individuals access services; expanded prevention and educational efforts—particularly aimed at teens, parents and other caretakers, and aging populations—to prevent the abuse of opioids and heroin and to promote treatment and recovery; expanded recovery support for students in high school or enrolled in institutions of higher learning;

and expanded resources to identify and treat incarcerated individuals suffering from addiction disorders promptly by collaborating with criminal justice stakeholders and by providing evidence-based treatment. CARA also expanded the availability of naloxone to law enforcement agencies and other first responders to help in the reversal of overdoses to save lives. CARA also reauthorizes a grant program for residential opioid addiction treatment of pregnant and postpartum women and their children and creates a pilot program for state substance abuse agencies to address identified gaps in the continuum of care, including non-residential treatment services.[14]

Are these laws effective?

In all three of these categories, the primary aim of the legislation in question is to decrease both the rates of addiction in and the negative impacts of addiction on both individuals and society at large, but in very different ways – either by limiting access to the substances themselves, or by punishing the use, sale, or possession, of them, or by facilitating the treatment and/or recovery of addicted individuals. But, of course, the most important question regarding these laws that we must ask is, do they work? That is, once laws such as these are implemented, we will want to know whether or not they have been effective in achieving their intended aims. Regarding laws in the first category, the intended aim is to decrease substance use via restricting access. Although there is disagreement over how best to measure the effectiveness of these laws, there is a general consensus that, on their own, these kinds of laws have not been enough to effectively address the substance abuse problem in this country (Crowley et al. 2017), and that, therefore other sorts of legislation is needed as well.

Regarding the laws in the second category, the intended aim is to reduce the use of addictive substances by punishing the individuals who use them. Regarding these laws, there is a growing consensus that they are not very effective (Edison 1978; Morse 2012) and, in some cases, can even be harmful. For example:

> The cost benefit critique of criminalization argues that such costs are not outweighed by the benefits because criminal law makes only a small dent in the use of drugs and because criminalization itself creates avoidable harms. The attempt to eradicate drug use by criminal prohibition cannot fully succeed because large numbers of people want recreational drugs for the pleasure or relief they provide, and it is widely recognized that the dangers of drugs are sometimes exaggerated. Given the powerful factors that motivate the desire for drugs, the criminal sanction appears ineffective. The criminal justice system cannot prosecute and imprison more than a tiny fraction of the enormous numbers of people involved in the illegal drug trade unless the justice system massively

diverts resources from other, undisputed criminal law needs and abandons civil liberties protection.

(Morse 2012)

Because of this, in recent years there has been a push for more decriminalization of addictive substances (Bretteville-Jensen 2009) and a greater use of funds for laws in the third category, where the intended aim of the legislation is not to punish addicts but rather to reduce both the rate and negative effects of substance addiction by providing recovery support to addicted individuals and their families. Yet, while many cite their support of laws in this category, they go on to argue that, while they are well-intended, they don't do enough:

> Despite these noteworthy increases in attention to and funding for substance use and addiction, especially in ways consistent with an evidence-based public health approach, prevention has received mostly lip service within the substance-specific funding streams. Initiatives that have been promoted within these laws primarily have focused either on adding the topic of prescription opioid misuse to existing drug prevention curricula, modifying clinical practice to reduce access to prescription opioids, or preventing opioid overdose deaths. While helpful and necessary, this approach is not sufficient to curb future drug epidemics, including the growing cases of stimulant misuse and addiction we are currently facing in the United States. Our country traditionally underinvests in prevention and tends to take a narrow, drug-specific approach that fails to address the root causes of substance use, build youth resilience, or adequately protect our nation from experiencing the next substance use and addiction crisis.[15]

Thus, we can see that while various laws directed toward solving the problem of substance abuse in the United States have been implemented widely, there remains controversy regarding their rates of success, as well as about how and whether they should be modified moving forward. It is important that we recognize this, both in this case and in others like it, because, of course, as we also saw in Chapter 4, there is no reason to enforce public policies that are not effective in achieving their aims. The goal of any legislation should be to improve conditions in the communities in which they take effect, and while science can certainly help us to do this, in order to justify policy interventions, we must appeal to extra-scientific values and considerations as well.

Conclusion

In this chapter I have aimed to show, via the example of laws governing addictive substances in the United States, that the process of enacting science-informed public policies requires several steps, beginning with defining the problem, followed next by a determination of whether or not the problem is worthy of a policy intervention, then by the proposal of a potential solution, next the communication of this solution to the relevant stakeholders, including both policymakers and the general public, and then finally, after implementing the solution, a determination of whether or not the intervention was, on the balance, effective and not harmful to the aims it was designed to achieve. Along the way, at each step in the policymaking process, scientists and other experts must be sure to acknowledge and communicate both any uncertainty in their data and/or predictions as well as any and all appeals to extra-scientific values. Doing these things allows non-scientist stakeholders to more effectively participate in the process, thereby facilitating better, and more effective, policies.

Notes

1 Although some (such as Morse 2012) argue that "it is possible to be a highly functioning addict who does not suffer or impose substantial negative consequences" (p. 264).
2 https://nida.nih.gov/publications/drugs-brainsbehavior-science-addiction/introduction
3 https://nida.nih.gov/publications/drugs-brains-behavior-science-addiction/introduction
4 https://drugabusestatistics.org/
5 https://nida.nih.gov/publications/drugs-brains-behavior-science-addiction/preface
6 (https://www.who.int/about/governance/constitution)
7 https://drugfree.org/article/is-addiction-a-disease/#Why%20some%20people%20say%20addiction%20is%20not%20a%20disease
8 https://www.ncbi.nlm.nih.gov/pmc/articles/PMC2715956/
9 https://www.ncbi.nlm.nih.gov/pmc/articles/PMC2715956/
10 https://www.asam.org/quality-care/definition-of-addiction/glossary-of-addiction
11 https://www.aaas.org/intersection-science-and-public-policy
12 https://ehs.usc.edu/research/cspc/chemicals/
13 https://www.samhsa.gov/about-us/who-we-are/laws-regulations
14 https://www.samhsa.gov/about-us/who-we-are/laws-regulations
15 https://www.healthaffairs.org/do/10.1377/forefront.20210607.239986/

Conclusion

As we have seen from the discussions in this book, the question of how to best apply scientific data to public policy is both a timely and enduring one. However, what often goes unrecognized in many texts on the topic is the fact that before we can weigh in on this important and practical question, we must first understand both what science *is* as well as what it can and cannot *do*. This, in turn, requires an understanding of the method and domain of science, as well as addressing the question of what counts as evidence toward a scientific theory (and who decides) in addition to questions about how to interpret scientific results, how to make sense of scientific dissent, how to communicate scientific data to non-scientist stakeholders, how to identify scientific experts, and, ultimately, how to apply scientific results to public policy in the context of a democratic society with a plurality of values, goals, and aims, without resorting to relativism. In particular, it is important for anyone who studies science, public health, or public policy more generally, to truly understand both what science is, and how it works, in order to be prepared to apply it.

To that end, the main goal of this book has been two-fold: the first, to explicate the methodology, domain, and aims of modern scientific inquiry, and the second, to show how an understanding of these things can enable us to effectively apply scientific research results to our public policies and practices. Thus the conceptual background material that is covered in the first part of the book should be recognized as vitally important to the overall project, because if we do not first understand what science is, how it works, what it aims for, or what its limitations are, we will be likely to either over or underestimate its predictive and/or explanatory power. This, in turn, will be likely to inhibit effective or meaningful application to our public policies. The idea is simple: to *apply* science, we must first *understand* it. Finally, we also saw via the examples that we examined in the book, which ranged from astrophysics to epidemiology to climate science, to laws governing addictive substances, that when applying data to policy, we need to

DOI: 10.4324/9781003311072-7

understand how to interpret, that is, how to make sense of, our data and results, as well as how to communicate these findings effectively. And to do this, we also need to understand the proper role of scientific experts in fostering this aim. When all of these things are done well, then we can reasonably expect our policies to be beneficial, well-informed, and effective in their aims. Thus, my hope is that what I have written here has been interesting, accessible (to both academics and non-academics alike), and ultimately motivating for those interested in the nature of science as well as in its potential for application to public policy, and that reading this book will have proven useful for those who will put eventually be the ones to put these principles into practice.

References

Abdel-Motleb, M. (2012). "The Neuropsychiatric Aspect of Addison's Disease." *Innovations in Clinical Neuroscience* 9: 34–36.

American Academy of Pediatrics, Task Force on Sudden Infant Death Syndrome (2005). "The Changing Concept of Sudden Infant Death Syndrome: Diagnostic Coding Shifts; Controversies Regarding the Sleeping Environment; and New Variables to Consider in Reducing Risk." *Pediatrics* 116 (5): 1245–1255.

Anderson, E. (2011). "Democracy, Public Policy, and Lay Assessments of Scientific Testimony." *Episteme* 8 (2): 144–164.

Ankeny, R. (2014). "The Overlooked Role of Cases in Casual Attribution in Medicine." *Philosophy of Science* 81 (5): 999–1011.

Appelbaum, P. S., L. H. Roth, C. W. Lidz, P. Benson, and W. Winslade (1987). "False Hopes and Best Data: Consent to Research and the Therapeutic Misconception." *The Hastings Center Report* 17 (2): 20–24.

Baron, J. (2012). "The Point of Normative Models in Judgment and Decision Making." *Frontiers in Psychology* 3: Article 577.

Bayer, R., and J. Colgrove (2002). "Science, Politics, and Ideology in the Campaign against Environmental Tobacco Smoke." *American Journal of Public Health* 92 (6): 949–954.

Beebe, J., and F. Dellsen (2020). "Scientific Realism in the Wild: An Empirical Study of Seven Sciences and History and Philosophy of Science." *Philosophy of Science* 87: 336–364.

Bendiscioli, S. (2019). "The Troubles with Peer Review for Allocating Research Funding." *EMBO Reports* 20 (12): e49472.

Bennett, M. (2020). "Should I Do as I'm Told? Trust, Experts, and Covid 19." *Kennedy Institute of Ethics Journal*. Special Issue. https://kiej.georgetown.edu/trust-experts-and-covid-19-special-issue/.

Bergen, N., Katherine Kirkby, Cecelia Vidal Fuertes, Anne Schlotheuber, Lisa Menning, Stephen Mac Feely, et al. (2023). "Global State of Education-related Inequality in COVID-19 Vaccine Coverage, Structural Barriers, Vaccine Hesitancy, and Vaccine Refusal: Findings from the Global COVID-19 Trends and Impact Survey." *The Lancet* 11 (2): e207–e217.

Berner, E. S., and M. L. Graber (2008). "Overconfidence as a Cause of Diagnostic Error in Medicine." *The American Journal of Medicine* 121 (Suppl. 5): S2–S23.

Blystone, R., and K. Blodgett (2006). "WWW: The Scientific Method." *CBE: Life Sciences Education* 5 (1): 7–11.

Brandt, A. (2012). "Inventing Conflicts of Interest: A History of Tobacco Industry Tactics." American *Journal of Public Health* 102 (1): 63–71.

Bretteville-Jensen, A. (2009). "To Legalize Or Not to Legalize? Economic Approaches to the Decriminalization of Drugs." *Substance Use & Misuse* 41 (4): 555–565.

Brownson, Ross, Chriqui, Jamie, and Katherine A. Stamatakis, (2009). "Understanding Evidence-Based Public Health Policy." *American Journal of Public Health* 99 (9): 1576–1583.

Cairney, P., and K. Oliver (2017). "Evidence-Based Policymaking Is Not Like Evidence-Based Medicine, So How Far Should You Go to Bridge the Divide between Evidence and Policy?" *Health Research Policy and Systems* 15 (1): 35.

Caruso, G., and O. Flanagan (2018). *Neuroexistentialism: Meaning, Morals, and Purpose in the Age of Neuroscience.* Oxford: Oxford University Press.

Cassam, Q. (2017). "Diagnostic Error, Overconfidence and Self-knowledge." *Palgrave Communications* 3: 17025.

Crowley, et al. (2017). "Prevention and Treatment of Substance Use Disorders Involving Illicit and Prescription Drugs: An American College of Physicians Position Paper." *Annals of Internal Medicine.* https://doi.org/10.7326/M16-2953.

Center on Addiction (2010). "Behind Bars II: Substance Abuse and America's Prison Population." February 2010. https://www.centeronaddiction.org/addiction-research/reports/behind-bars-ii-substance-abuse-and-america's-prison-population.

Centers for Disease Control and Prevention (2023). "What Is Polio?" https://www.cdc.gov/polio/what-is-polio/index.htm.

——— (2007). "Reduced Secondhand Smoke Exposure after Implementation of a Comprehensive Statewide Smoking Ban–New York, June 26, 2003–June 30, 2004." *Morbidity and Mortality Weekly Report* 56 (28): 705–708 [accessed 1 May 2014].

Coady, D., and R. Corry (2013). *The Climate Change Debate: An Epistemic and Ethical Enquiry.* New York: Palgrave Macmillan.

Cowley, C. (2012). "Expertise, Wisdom and Moral Philosophers: A Response to Gesang." *Bioethics* 26: 337–342.

Croce, M. (2019). "On What It Takes to Be an Expert." *Philosophical Quarterly* 69 (274): 1–21.

Crosskerry, P., and G. Norman (2008). "Overconfidence in clinical Decision Making." *The American Journal of Medicine* 121 (5): 224–S29.

Daniel, M., S. Khandelwal, S. A. Santen, M. Malone, and P. Croskerry (2017). "Cognitive Debiasing Strategies for the Emergency Department." *AEM Education and Training* 1 (1): 41–42.

Demicheli, et al. (2012). "Vaccines for Measles, Mumps and Rubella in Children." *Cochrane Database of Systematic Reviews* 2: CD004407. https://doi.org/10.1002/14651858.CD004407.pub3.

Douglas, H. (2013). "The Value of Cognitive Values." *Philosophy of Science* 80 (5): 796–806.

——— (2009). *Science, Policy and the Value-Free Ideal.* Pittsburgh, PA: University of Pittsburgh Press.

Edison "The Drug Laws: Are They Effective and Safe?" *JAMA* 239 (24): 2578–2583.

Elliott, K. (2021). "Studies in History and Philosophy of Science 88." *Philosophy of Science* 78: 303–324.

———— (2019). "Managing Value-Laden Judgements in Regulatory Science and Risk Assessment." *Efsa Journal* 17(Suppl 1): e170709. doi: 10.2903/j.efsa.2019. e170709.

———— (2017). *A Tapestry of Values*. New York: Oxford University Press.

———— (2011). "Direct and Indirect Roles for Values in Science." *EFSA Journal* 17 (Suppl. 1): e170709.

Elliott, K., and D. B. Resnik (2014). "Science, Policy, and the Transparency of Values." *Environmental Health Perspectives* 122: 647–650.

Emanuel, E. J. (2003). *Ethical and Regulatory Aspects of Clinical Research: Readings and Commentary*. Baltimore, MD: Johns Hopkins Press.

Emanuel, E. J., D. Wendler, and C. Grady (2000). "What Makes Clinical Research Ethical?" *JAMA* 283 (20): 2701–2711.

Ferguson et. al. (2020). "Report 9: Impact of non-pharmaceutical interventions (NPIs) to reduce COVID-19 mortality and healthcare demand" https://www.imperial.ac.uk/media/imperial-college/medicine/sph/ide/gida-fellowships/Imperial-College-COVID19-NPI-modelling-16-03-2020.pdf

Fischer, E. (1996). "The Public Misunderstanding of Science." *Interdisciplinary Science Reviews* 21 (2): 110–116.

Fischoff, B. and A.L. Davis. (2014). "Communicating Scientific Uncertainty." *Proceedings of the National Academy of Sciences of the United States of America*, 11 (Suppl 4): 13664–13671.

Funk, C. (2017). "Real Numbers: Mixed Messages about Public Trust in Science." *Issues in Science and Technology* 34. https://www.pewresearch.org/science/2017/12/08/mixed-messages-about-public-trust-in-science/

Gallo, S., J. Sullivan, and S. Glisson (2016). "The Influence of Peer Reviewer Expertise on the Evaluation of Research Funding Applications." *PLOS One*. https://doi.org/10.1371/journal.pone.0165147.

Goldberg, A. E. (2020). "The (in)Significance of the Addiction Debate." *Neuroethics* 13: 311–324. https://doi.org/10.1007/s12152-019-09424-5.

Goldenberg, M. (2021). *Vaccine Hesitancy: Public Trust, Expertise, and the War on Science*. Pittsburgh: University of Pittsburgh Press.

———— (2016). "Public Misunderstanding of Science? Reframing the Problem of Vaccine Hesitancy." *Perspectives on Science* 24(5): 552–581.

Goldman, A. (2018). "Expertise." *Topoi* 37(1): 3–10.

———— (2001). "Experts: Which Ones Should You Trust?" *Philosophy and Phenomenological Research* 63: 18–212.

Goldstein, R. Z., and N. D. Volkow (2011). "Dysfunction of the Prefrontal Cortex in Addiction: Neuroimaging Findings and Clinical Implications." *Nature Reviews Neuroscience* 12 (11): 652–669. https://doi.org/10.1038/nrn3119.

Gorwood, P., Y. Strat, and N. Ramoz (2019). "A Genetic Framework for Addiction." In *The Routledge Handbook of Philosophy and Science of Addiction*, edited by H. Pickard and S. H. Ahmed, 275–285. New York: Routledge.

Graham, A. P., B. Butler, L. Kogan, L., P. Palmer, and V. Strelnitski (2000). "Water Maser Emission from Comets." (with Butler, Kogan, Palmer and Strelnitski) *Astronomical Journal* 119: 2465–2471.

Grier, J. (2017). "We Used Terrible Science to Justify Smoking Bans." *Slate Magazine*. https://slate.com/technology/2017/02/secondhand-smoke-isnt-as-bad-as-we-thought.html.

Hahn, E. J. (2010). "Smokefree Legislation: A Review of Health and Economic Outcomes Research." *American Journal of Preventive Medicine* 39 (6 Suppl. 1): S66–S76.

Havstad, J., and M. Brown (2017). "Inductive Risk, Deferred Decisions, and Climate Science Advising." In *Exploring Inductive Risk: Case Studies of Values in Science*, edited by K. C. Elliott and T. Richards. Oxford: Oxford University Press.

Haw, S. and L. Gruer (2007). "Changes in Exposure of Adult Non-Smokers to Secondhand Smoke after Implementation of Smoke-Free Legislation in Scotland: National Cross Sectional Survey." *BMJ* 335 (7619): 549.

Henden, E. (2019). "Addiction as a Disorder of Self Control." In *The Routledge Handbook of Philosophy and Science of Addiction*, edited by H. Pickard and S. H. Ahmed, 45–53. New York: Routledge.

Holman, B. (2015). "Most Sugar Pills Are Not Placebos." *Philosophy of Science* 82: 1330–1343.

Howick, J. (2011). *The Philosophy of Evidence-Based Medicine*. Oxford: Wiley-Blackwell.

Huang, C.-Y., and S.-F. Chen (2006). "Reducing the Risk of Sudden Infant Death Syndrome through Safe Sleep Practices." *National Library of Medicine Review*. https://pubmed.ncbi.nlm.nih.gov/16874597/.

Huber, G. (1975). "Smoking and Nonsmokers—What Is the Issue?" *The New England Journal of Medicine* 292: 858–859.

Hunter, J., and P. Nedelisky (2018). *Science and the Good: The Tragic Quest for the Foundations of Morality*. New Haven, CT: Yale University Press.

Institute of Medicine (US) Committee on Secondhand Smoke Exposure and Acute Coronary Events (2010). *Secondhand Smoke Exposure and Cardiovascular Effects: Making Sense of the Evidence*. Washington, DC: National Academies Press.

Jamieson, D. (1996). "Scientific Uncertainty and the Political Process." *Annals of the American Academy of Political and Social Science* 545: 35–43.

Jasanoff, S. (1990). *The Fifth Branch: Science Advisors as Policymakers*. Cambridge, MA: Harvard University Press.

John, S. (2019). "Science, Truth and Dictatorship: Wishful Thinking Or Wishful Speaking?" *Studies in History and Philosophy of Science* 78: 64–72.

——— (2018). "Epistemic Trust and the Ethics of Science Communication: Against Transparency, Openness, Sincerity and Honesty." *Social Epistemology* 2: 75–87.

Jukola, S. (2022). "On the Evidentiary Standards for Nutrition Advice." *Studies in History and Philosophy of Science Part C. Studies in Biology and Biomedical Sciences* 73: 1–9.

Kahan, D. M., E. Peters, M. Wittlin, P. Slovic, L. L. Ouellette, D. Braman, and G. Mandel (2012). "The Polarizing Impact of Science Literacy and Numeracy on Perceived Climate Change Risks." *Nature Climate Change* 2 (10): 732–735. https://doi.org/10.1038/nclimate1547.

Kahneman, D., and S. Frederick (2005). "A Model of Heuristic Judgment." In *The Cambridge Handbook of Thinking and Reasoning*, edited by K. J. Holyoak and R. G. Morrison, 267–293. Cambridge: Cambridge University Press.

Kennedy, A. (2021). *Diagnosis: A Guide for Medical Trainees*. New York: Oxford University Press.

———— (2015a) "Evaluating Diagnostic Tests," *Journal of Evaluation in Clinical Practice.*

———— (2015b) "The Vaccine Debate: Where Do We Go from Here?" *Prindle Post.*

Kennedy, A., and B. Cwik (2021). "Diagnostic Justice: Testing for COVID-19." *European Journal of Analytical Philosophy.*

Kennedy, A., and S. Malanowski (2018). "Mechanistic Reasoning and Informed Consent." *Bioethics* 33 (1): 162–168.

Kiene, H., H. Hamre, and G. Kienle (2013). "In Support of Clinical Case Reports: A System of Causality Assessment." *Global Advances in Health and Medicine* 2 (2): 64–75.

Kim, J. (1988). "What Is Naturalized Epistemology?" In *Philosophical Perspectives 2, Epistemology*, edited by J. Tomberlin, 381–405. Atascadero, CA: Ridgeview Publishing Co.

Kitcher, P. (2011). *Science in a Democratic Society*. New York: Prometheus Books.

Kuhn, T. S. (1962). *The Structure of Scientific Revolutions*. Chicago, IL: Chicago University Press.

Larrick, R.P. and J. B. Soll (2012). "Combining the Intuitive and Analytic Mind." *Paper presented at the Behavioral Decision Research in Management*, Boulder, CO.

Lander, L., J. Howsare, and M. Byrne (2013). "The Impact of Substance Use Disorders on Families and Children: From Theory to Practice." *Social Work in Public Health* 28: 194–205.

Lekka-Kowalik, A. (2010). "Why Science Cannot Be Value Free." *Science and Engineering Ethics* 16 (1): 33–41.

Levy, N. (2019). "The Belief Oscillation Hypothesis." In *The Routledge Handbook of Philosophy and Science of Addiction*, edited by Hanna Pickard and Serge Ahmed, 54–63. https://www.routledge.com/The-Routledge-Handbook-of-Philosophy-and-Science-of-Addiction/Pickard-Ahmed/p/book/9780367571504

Mackay, D. F., M. O. Irfan, S. Haw, and J. P. Pell (2010). "Meta-Analysis of the Effect of Comprehensive Smoke-Free Legislation on Acute Coronary Events." *Heart* 96: 1525–1530.

Malanowski, S., A. Kennedy, and N. Baima (2022). "Science Shame and Trust." In *Trust in Science*, edited by Kennedy, Ashley Graham, with Sarah Malanowski, and Nicholas Baima. Springer. (forthcoming)

Marti, J., and J. Schlapfer (2014). "The Economic Impact of Swiss Smoking Bans on the Hospitality Sector." Economics Letters 124 (1): 136–139.

Mason, S. (2020). "Climate Science Denial as Willful Hermeneutical Ignorance." *Social Epistemology* 34: 469–477.

Mathews, C., L. McGuire, A. Joy, F. Law, M. Winterbottom, A. Rutland , M. Drews, A. J. Hoffman, K. L. Mulvey, and A. Hartstone-Rose (2021). "Assessing Adolescents' Critical Health Literacy: How Is Trust in Government Leadership Associated with Knowledge of COVID-19?" *PLOS.* https://doi.org/10.1371/journal.pone.0259523

McCormick, J.B. (2018). "How Should a Research Ethicist Combat False Beliefs and Therapeutic Misconception Risk in Biomedical Research?" *AMA Journal of Ethics* 20 (11): E1100–1106.

Meldrum, M. (1998). ""A Calculated Risk": The Salk Polio Vaccine Field Trials of 1954." *BMJ* 317 (7167): 1233–1236.

Mercuri, M. (2020). "Just Follow the Science: A Government Response to a Pandemic." *Journal of Evaluation in Clinical Practice*. https://onlinelibrary.wiley.com/doi/full/10.1111/jep.13491.

Miller, F. G., and D. L. Rosenstein (2003). "The Therapeutic Orientation to Clinical Trials." *New England Journal of Medicine* 348 (14): 1383–1386.

Moon, R. (2017). "SIDS and Other Sleep-Related Infant Deaths: Expansion of Recommendations for a Safe Infant Sleeping Environment." *Pediatrics* 128 (5): e1341–e1367.

Morse, S. J. (2012). "Legal Regulation of Addictive Substances and Addiction." In *Addiction Neuroethics: The Ethics of Addiction Neuroscience Research and Treatment*, edited by A. Carter, W. Hall and J. Illes, 261–276. San Diego, CA: Academic Press. https://doi.org/10.1016/B978-0-12-385973-0.00014-4.

Morewedge, C., H. Yoon, I. Scopelliti, J. H. Symborski, H. Korris, and K. Kassam (2015). "Debiasing Decisions: Improved Decision Making with a Single Training Intervention." *Policy Insights from the Behavioral and Brain Sciences* 2 (1): 129–140.

National Human Genome Research Institute (2005). "The Use of Racial, Ethnic, and Ancestral Categories in Human Genetics Research." *American Journal of Human Genetics* 77 (4): 519–532.

Oreskes, N., and E. Conway (2010). *Merchants of Doubt*. New York: Bloomsbury.

Peres, J. (2013). "No Clear Link between Passive Smoking and Lung Cancer." *Journal of the National Cancer Institute* 105 (24): 1844–1846.

Pickard, H. (2019). "The Puzzle of Addiction." In *The Routledge Handbook of Philosophy and Science of Addiction*, edited by H. Pickard and S. H. Ahmed, 9–22. New York: Routledge.

Pinto, M., and D. Hicks (2019). "Legitimizing Values in Regulatory Science." *Environmental Health Perspectives* 127 (3). doi: 10.1289/EHP3317.

Potochnick, A. (2017). *Idealization and the Aims of Science*. Chicago, IL: University of Chicago Press.

Prasad, Vinay (2020). *Op-Ed: What Does 'Follow the Science' Mean, Anyway? — Science is a Tool, not a Prescription for Policy on COVID-19*. https://www.medpagetoday.com/opinion/vinay-prasad/89856#:~:text=Policy%20is%20a%20human%20endeavor%20that%20combines%20science,tell%20you%20whether%20to%20open%20or%20close%20schools.

Psillos, S. (1999). *Scientific Realism: How Science Tracks Truth*. London: Routledge.

Racine, E., S. Kusch, M.A. Cascio, et al. (2021). "Making Autonomy an Instrument: A Pragmatist Account of Contextualized Autonomy." *Humanities and Social Sciences Communications* 8: 139. https://doi.org/10.1057/s41599-021-00811-z

Reiss, J. (2019). "Expertise, Agreement, and the Nature of Social Scientific Facts Or: Against Epistocracy." *Social Epistemology* 33 (2): 183–192.

Rosenberg, N., J.K Pritchard, H.M. Weber, K. K. Cann and I.A. Kidd (2002). "Genetic Structure of Human Populations." *Science* 298: 2381–2385.

Rudner, R. (1953). "The Scientist Qua Scientist Makes Value Judgments." *Philosophy of Science* 20 (1): 1–6.

Schmidt, J. (ed.) (1996). *What Is Enlightenment? Eighteenth-Century Answers and Twentieth-Century Questions.* Berkeley: University of California Press.

Schroeder, S. A. (2019). "Democratic Values: A Better Foundation for Public Trust in Science." *The British Journal for the Philosophy of Science* 72(2). https://www.journals.uchicago.edu/doi/abs/10.1093/bjps/axz023

Schwab, A. (2012). "Epistemic Humility and Medical Practice: Translating Epistemic Categories into Ethical Obligations." *Journal of Medicine and Philosophy* 37: 28–48.Sehon, Scott and Donald Stanley. (2003). "A Philosophical Analysis of the Evidence-based Medicine Debate." *BMC Health Services Research* 3(1): 14. doi: 10.1186/1472-6963-3-14.

Scientific uncertainty. Nat. Clim. Chang. 9, 797 (2019). https://doi.org/10.1038

Shahar, Eyal (1997). "A Popperian Perspective on the Term "evidence-based medicine"". *Journal of Evaluation in Clinical Practice* 3 (2): 109–16.

Simon, J. (2017). "Realism and Constructivism in Medicine." In *Routledge Companion to Philosophy of Medicine*, edited by M. Solomon, J. R. Simon and H. Kincaid. Philadelphia: Routledge.

Simon, J., et al. (2021). https://bmcpublichealth.biomedcentral.com/articles/10.1186/s12889-021-10351-5.

Singer, S. F. (1991). "Global Warming: Do We Know Enough to Act?" In *Environmental Protection: Regulating for Results*, edited by K. Chilton and M. Warren, 29–49. Boulder, CO: Westview Press.

Sinott-Armstrong, W., and J. S. Summers (2019). "Defining Addiction: A Pragmatic Perspective." In *The Routledge Handbook of Philosophy and Science of Addiction*, edited by H. Pickard and S. H. Ahmed, 123–131. New York: Routledge.

Smith, R. (2006). "Peer Review: A Flawed Process at the Heart of Science and Journals." *Journal of the Royal Society of Medicine* 99 (4): 178–182.

Soll, J , K. Milkman, and J. Payne (2016). "A User's Guide to De-biasing." https://papers.ssrn.com/sol3/papers.cfm?abstract_id=2455986

Solomon, M. (2001). *Social Empiricism.* Boston: MIT Press.

Sperber, D., F. Clément, C. Heintz, O. Mascaro, H. Mercier, G. Origgi, and D. Wilson (2010). "Epistemic Vigilance." *Mind & Language* 25 (4): 359–393. BMC Public Health, Article number: 314.

Stanev, R. (2017). "Inductive risk and Values in Composite Outcome Measures." In *Exploring Inductive Risk: Case Studies of Values in Science*, edited by K. Elliott and T. Richards, 171–191. New York: Oxford University Press

Stanford, K. (2007). "Risk." *Stanford Encyclopedia of Philosophy.* https://plato.stanford.edu/entries/risk/.

Stone, D., C. Kerr, E. Jacobson, L. Conboy, and T. Kaptchuk (2005). "Patient Expectations in Placebo-Controlled Randomized Clinical Trials." *Journal of Evaluation in Clinical Practice* 11 (1): 77–84.

Tetlock, Philip. 2005. *Expert Political Judgment: How Good Is It? How Can We Know?* Princeton, NJ: Princeton University Press.

Thomson, G., and N. Wilson (2006). "One Year of Smoke-Free Bars and Restaurants in New Zealand: Impacts and Responses." *BMC Public Health* 6: 64.

Tufeki, Z. (2020). "Why Telling People They Don't Need Face Masks Backfired." *New York Times*. https://www.nytimes.com/2020/03/17/opinion/coronavirus-face-masks.html.

Tukey, J. W. (2014). *The Health Consequences of Smoking—50 Years of Progress: A Report of the Surgeon General*. Atlanta, GA: U.S. Department of Health and Human Services, Centers for Disease Control and Prevention, National Center for Chronic Disease Prevention and Health Promotion, Office on Smoking and Health.

U.S. Department of Health and Human Services (2006). *The Health Consequences of Involuntary Exposure to Tobacco Smoke: A Report of the Surgeon General*. Atlanta, GA: U.S. Department of Health and Human Services, Centers for Disease Control and Prevention, Coordinating Center for Health Promotion, National Center for Chronic Disease Prevention and Health Promotion, Office on Smoking and Health.

van Fraassen, B. (1980). *The Scientific Image*. Oxford: Oxford University Press.

Wakefield, J. (2019). "The Harmful Dysfunctional Analysis of Addiction." In *The Routledge Handbook of Philosophy and Science of Addiction*, edited by H. Pickard and S. H. Ahmed, 90–101. New York: Routledge.

Walton, D. (1989). "Reasoned Use of Expertise in Argumentation." *Argumentation* 3: 59–73.

Ware, J. H., and M. D. Epstein (1985). "Comments on "Extracorporeal Circulation in Neonatal Respiratory Failure: A Prospective Randomized Study" by R. H. Bartlett et al." *Pediatrics* 76: 849–851.

Watson, G. (2004). "Disordered Appetites: Addiction, Compulsion, and Dependence." In *Agency and Answerability: Selected Essays* (online ed.). Oxford: Oxford Academic. https://doi.org/10.1093/acprof:oso/9780199272273.003.0004 [accessed 13 August 2022].

West, P. "An Egalitarian Evaluation of Lockdown." Presented at Durham University, January 26, 2022.

Winsberg, E., Jason Brennan, and Chris W. Surprenant (2020). "How Government Leaders Violated Their Epistemic Duties during the SARS-CoV-2 Crisis." *Kennedy Institute of Ethics Journal* 30 (3–4): 215–242.

Worrall, J. (2004). "Why There's No Cause to Randomize." Centre for Philosophy of Natural and Social Science Causality: Metaphysics and Methods Technical Report 24/04.

Index

Note: *Italic* page numbers refer to figures and page numbers followed by "n" denote endnotes.